Technical Communication Quarterly

Association of Teachers of Technical Writing
http://www.attw.org

Editors

Charlotte Thralls	Utah State University
Mark Zachry	Utah State University
Book Reviews: Tracy Bridgeford	University of Nebraska at Omaha
Special Issues: Sherry Burgus Little	San Diego State University
Richard Johnson-Sheehan	University of New Mexico

Editorial Advisory Board

Paul V. Anderson	Miami University, Ohio
Stephen A. Bernhardt	University of Delaware
Deborah S. Bosley	University of North Carolina, Charlotte
Rebecca Burnett	Iowa State University
Barbara Couture	Washington State University
Sherry Burgus Little	San Diego State University
Katherine Staples	Austin Community College
Elizabeth Tebeaux	Texas A&M University
James P. Zappen	Rensselaer Polytechnic Institute

Managing Editor

Philip Parisi	Utah State University

ATTW Executive Committee Officers

President: Jo Allen	North Carolina State University
Vice President: Bill Karis	Clarkson University
Secretary: Brenda Sims	University of North Texas
Treasurer: Ann Blakeslee	Eastern Michigan University

Members at Large

Rebecca Burnett	Iowa State University
Teresa Kynell-Hunt	Northern Michigan University
Dorothy Winsor	Iowa State University

Past Presidents

Carolyn Rude	Virginia Tech

Correspondence concerning ATTW should be sent to Jo Allen, 128 Leazar, Campus Box 7105, North Carolina State University, Raleigh, NC 27695.

TCQ Reviewers

TCQ relies on the expertise of our reviewers not only to select articles for publication but also to help authors see possibilities for development. We thank them for their contributions to the quality of the journal.

Paul V. Anderson	Miami University, Ohio
Carol Berkenkotter	University of Minnesota
Stephen Bernhardt	University of Delaware
Ann M. Blakeslee	Eastern Michigan University
Lee Brasseur	Illinois State University
Davida Charney	University of Texas
Mary B.Coney	University of Washington
Jennie Dautermann	Miami University, Ohio
Mary Beth Debs	University of Cincinnati
Paul Dombrowski	University of Central Florida
Sam Dragga	Texas Tech University
Brenton Faber	Clarkson University
David Farkas	University of Washington
Alexander Friedlander	Drexel University
Karen Griggs	Indiana University-Purdue University
Barbara Heifferon	Clemson University
Jim Henry	George Mason University
Robert R. Johnson	Michigan Technological University
Jim Kalmbach	Illinois State University
Bill Karis	Clarkson University
Steven B. Katz	North Carolina State University
Charles Kostelnick	Iowa State University
Carol Lipson	Syracuse University
Bernadette Longo	University of Minnesota
Barbara Mirel	University of Michigan
Michael G. Moran	University of Georgia
Meg Morgan	University of North Carolina, Charlotte
Roger Munger	Boise State University
Kenneth Rainey	Southern Polytechnic State University
Frances Ranney	Wayne State University
Fred Reynolds	City College of CUNY
Lilita Rodman	University of British Columbia
Mark Rollins	Ohio University
Carolyn D. Rude	Virginia Tech
Beverly Sauer	John Hopkins University
Gerald Savage	Illinois State University
Catherine F. Schryer	University of Waterloo
Blake Scott	University of Central Florida
Stuart A. Selber	Pennsylvania State University
Cynthia L. Selfe	Michigan Technological University
Jack Selzer	Pennsylvania State University
Brenda R. Sims	University of North Texas
Herb Smith	Southern Polytechnic State University
Bruce Southard	East Carolina University
Rachel Spilka	University of Wisconsin-Milwaukee
Sherry Southard	East Carolina University
Clay Spinuzzi	University of Texas at Austin
Dale Sullivan	North Dakota State University
Elizabeth Tebeaux	Texas A&M University
Emily A. Thrush	University of Memphis
Linda Van Buskirk	Cornell University
James P. Zappen	Rensselaer Polytechnic Institute

Technical Communication Quarterly

Volume 13, Number 1   Winter 2004

Guest Editor's Introduction

Carolyn Rude
Virginia Tech

Several recent developments in the history of the Association of Teachers of Technical Writing (ATTW) and *Technical Communication Quarterly (TCQ)* invite the essays in this issue on the state of technical communication in its academic context. ATTW, organized in 1973, celebrated its thirtieth anniversary; this issue of *TCQ* welcomes new editors Charlotte Thralls and Mark Zachry, both from Utah State University; and this is the first issue of *TCQ* published by Lawrence Erlbaum Associates, Inc. (LEA). In this issue, the ATTW Executive Committee has taken the opportunity that change offers for some self-study and reflection on the technical communication field and the role academics plays in it. This reflection should help both the leaders of the Association and of academics in general to develop a vision and to plan for the future.

As the article in this issue by Don Cunningham records, ATTW and *TCQ* began when faculty from several colleges and universities joined together with the goal of improving the quality of teaching of technical communication. Like all good technical communicators, they defined a need and collaborated on ways to meet it. They organized and started a journal designed to support faculty who found themselves thrust into teaching technical writing without much preparation or materials. Their groundwork has enabled the next generation of faculty to build on a strong foundation. The body of knowledge and support materials on teaching have evolved substantially in sophistication and quantity. Research continues to develop on best practices of writing in the workplace and on the cultural and rhetorical aspects of technical communication. ATTW encourages such development by providing forums for discussion and exchange. In addition, ATTW remains a collegial organization with ongoing traditions of collaboration, generosity, and goodwill. The organization has developed very much in focus with the vision of its founders. As times and needs change, however, visions must be reexamined and refocused, and the articles in this issue of *TCQ* facilitate that goal.

Our new editors were selected by the ATTW Executive Committee from a review of proposals submitted by several institutions. Charlotte Thralls, along with Nancy Roundy Blyler, was a founding editor of the *Journal of Business and Tech-*

nical Communication, and the expertise of these co-editors and their good management helped make *JBTC* a flagship journal from the start. Charie brings that experience to *TCQ*, along with a distinguished record of teaching and scholarship. Her professional interests are in the application of contemporary rhetorical theory to business and technical communication, with an emphasis on cultural studies and organizational communication. Mark Zachry joined the faculty at Utah State University in 1998, after earning his PhD from Iowa State University, and he now directs the program in professional and technical writing at Utah State. His teaching interests are rhetorical theory, information architecture, and usability. His research focuses on the sociohistorical aspects of professional communication, genre theory, and narrative. We welcome Charie and Mark to the editorship and wish them well as they pursue their vision for the journal.

At the same time, we thank with great appreciation our editors since 1991, Mary Lay and Billie Wahlstrom. Mary and Billie assumed editing responsibilities at the point when the journal's name changed from *The Technical Writing Teacher* to *Technical Communication Quarterly*; with the name change came new and ambitious goals, as Mary chronicles in her article in this issue. Under their guidance, the journal has matured to become a leading journal for research, theory, and pedagogy in technical communication. We also thank the University of Minnesota, which has hosted and supported the journal since 1984, when Victoria Mikelonis became its editor.

As LEA becomes the publisher of *TCQ*, the journal remains a mission of the ATTW. The editors, guided by the journal's advisory board and the ATTW Executive Committee, will continue to develop content and edit the journal, but LEA assumes responsibility for the business side of the journal: production, marketing, and ATTW membership renewals. A subscription to the journal will continue to be a benefit of membership, and members will be able to renew their memberships online at the LEA Web site by subscribing to the journal. LEA will provide online copies of the journal to members. ATTW is pleased about this association with LEA for more than new conveniences to members. Our relationship should result in an enhanced presence for ATTW in writing studies because of LEA's contacts with other professional associations and its publication of journals and books in rhetoric and writing. The relationship also frees the ATTW volunteers from some of the business of the association, such as membership renewals, so that they can focus more on its academic missions, especially developing the content of the journal, organizing the annual conference, and developing the book series and Web site. Members of the ATTW Executive Committee hope to provide better services to members with the help of LEA.

Most of the articles in this issue have been written by members of the Executive Committee and their collaborators or by others who have been identified as leaders in a particular subject area. The Winter 2004 issue focuses on the work of the Association as it helps to guide the evolution of the field, including a description of its members, reflections on the journal and its history, assessment of student learning, research in the field, and the academic job market. This issue includes two articles based on surveys of faculty in technical communication, one conducted by David Dayton and Steve Bernhardt of ATTW members, and one conducted by the Society for Technical Communication (STC) that focused on academic salaries and is reported by Sandi Harner. Dayton and Bernhardt constructed a profile of members, including their participation in ATTW activities and evaluation of ATTW's benefits, interests in *TCQ* topics, and hopes for the future. Harner discusses the results of the first STC salary survey of academics and reports on issues of job satisfaction and faculty benefits as well. Carolyn Rude and Kelli Cargile Cook have studied the academic job market in 2002–2003 to answer recurring questions about the demand for faculty in comparison with the supply of graduates of doctoral programs in the field and to describe the different types of academic jobs for which graduate students might prepare themselves. Jo Allen describes the values and methods of assessment based on student writing and learning outcomes. This method of assessment, based on criteria that faculty in the field establish, may provide better information than assessment methods and criteria imposed by college administrations. Ann Blakeslee and Rachel Spilka assess strengths and weaknesses of research in the field and propose a plan of action to develop our research. Mary Lay, who has co-edited *TCQ* for thirteen years, reflects on what constitutes a publishable article and the process of manuscript development and selection. Don Cunningham, the founding editor of ATTW's journal, first named *The Technical Writing Teacher,* recalls the early years of the association and its journal. Tracy Bridgeford's review of the collection edited by Rachel Spilka and Barbara Mirel (*Reshaping Technical Communication*) is, like the book itself, a reflection on the state of the profession.

The ATTW Executive Committee hopes that the information in this issue will benefit members at every level: members who would like to see how their own situations and interests compare with those of colleagues at other colleges and universities; faculty who are interested in the teaching and research issues of the field; graduate students who have opportunities to prepare themselves strategically to enter the field as faculty; program administrators and department chairs who are interested in the potential for development, components of which include assessment and faculty hiring and retention; potential authors of *TCQ* articles for whom knowing the criteria and procedures of manuscript review and selection will enhance their chances of publication; all members who welcome the sense of belonging to an organization with a history and traditions; and the leaders of ATTW, who can plan the organization's initiatives based on the expressed interests of the members they serve.

Carolyn Rude
Virginia Tech, Blacksburg
Past President, ATTW

Charlotte Thralls, Co-Editor

Charlotte Thralls—better known as Charie in ATTW circles—has been an active participant in the field of technical communication for over twenty-five years. Currently, she is a full professor in the Department of English at Utah State University, where she teaches in the undergraduate technical communication program, as well as in two graduate programs: Technical Communication and the Theory and Practice of Writing. Prior to this position, she enjoyed a long and satisfying career (twenty-two years) at Iowa State University, where she worked with colleagues to create the undergraduate, master's, and doctoral programs in Rhetoric and Professional Communication.

With a primary research interest in rhetorical and cultural theories, Charie has developed a variety of courses, ranging from rhetorical theory and analysis to the history of rhetoric, the rhetoric of science, professional communication theory and research, cultural studies, the teaching of technical communication, and propaganda analysis. She has also served as a coordinator of internships in technical communication and as a director of graduate studies.

This emphasis on contemporary theory—its implications for technical communication pedagogy and the study of organizational communication—is reflected in many of her publications, including *Professional Communication: The Social Perspective* (co-edited with Nancy Blyler), and an upcoming collection, *The Cultural Turn: Perspectives on Communicative Practices in Workplaces and the Professions* (co-edited with Mark Zachry).

In addition to her teaching and research, Charie has maintained an active record of service to the profession. For example, with Nancy Blyler, she cofounded and edited the *Journal of Business and Technical Communication*. She also has served on numerous national committees, including current membership on the National Council of Technical English (NCTE) Committee on Scientific and Technical Communication, the Publications Board for the Association for Business Communication, and several journal editorial/review boards.

Charie's teaching and research have been recognized with several awards, such as the Louis Thompson Award for Distinguished Undergraduate Teaching at Iowa State University, the NCTE award for Best Collection in Scientific and Technical Communication, and the Alpha Kappa Psi Award in Business Communication. Most recently, Charie was named the 2001 Outstanding Researcher by the Association for Business Communication and the 2003 Humanist (Researcher) of the Year by the College of Humanities, Arts, and Social Sciences at Utah State University.

Mark Zachry, Co-Editor

Mark Zachry spent seven years as a researcher and editor in the aerospace and computer industries before making the transition to higher education. Since 1999, he has served as the chair of the Professional & Technical Writing program at Utah State University, where he teaches a variety of undergraduate and graduate courses. He also serves as Director of the Usability Research and Evaluation Lab at the university, working with faculty, staff, students, and workplace professionals from a broad range of areas to design, conduct, and report research.

Mark's own research focuses on understanding the formation and transformation of communicative practices in human organizations, ranging from manufacturing centers to online classes. His work has appeared in the field's major journals and in edited collections including *Narrative and Professional Communication* and the forthcoming *Online Education: Global Questions, Local Answers*. In addition, he has a forthcoming collection with Charlotte Thralls, *The Cultural Turn: Perspectives on Communicative Practices in Workplaces and the Professions*.

The courses Mark has developed and taught at Utah State include technology and the writer, online help development, content management systems, professional writing capstone, modern rhetorical theory, genre theory, usability studies and human factors in professional communication, as well as an introduction to technical communication. His classes range from face-to-face instruction to completely online courses that serve graduate students around the globe. In addition to his teaching, Mark is an active consultant, working with companies in the service and technology industries.

Mark serves on the editorial/advisory boards for several journals and on the research committees for ATTW and the Association for Business Communication. For the past five years, he has served as a contributing editor to *Business Communication Quarterly*.

Mark's work as a teacher and researcher has been recognized with awards from both Iowa State University and Utah State University. In 2003, Mark was named Teacher of the Year by the College of Humanities, Arts, and Social Sciences at Utah State University.

A Message from the New Editors

Charlotte Thralls and Mark Zachry

Utah State University

As a scholarly journal with nearly a thirty-year history of publishing articles on the teaching, study, and practice of communication in academic, scientific, technical, governmental, and business/industrial contexts, *Technical Communication Quarterly* has clearly established itself as a premiere venue for scholarship in our field. We are honored at this juncture in the journal's history that we have been selected to serve as its editors.

In this issue, Don Cunningham and Mary Lay, former editors, provide a fascinating chronicle of key moments in the journal's history. These reflections, of course, have important documentary value, but, equally important, they provide insight into the enormous contributions that others have made to ensure the journal's success. As the new *TCQ* editors, we want to acknowledge our indebtedness to these predecessors. In particular, we want to thank Don Cunningham, founding editor, and his colleagues for their prescience and courage in launching the journal. We also want to recognize the University of Minnesota, which for an incredible twenty years provided an institutional home for the journal. During that time, Victoria Mikelonis, Mary Lay, and Billie Wahlstrom served terms as exceptional editors, not only building and transforming the journal but also overseeing all the complex aspects of journal operation. Charie offers a special thank you to Victoria Mikelonis, editor from 1984 to 1990, for sharing her expertise with Charie and Nancy Blyler during the start-up of *The Journal of Business and Technical Communication* (*JBTC*), helping that potentially competing journal become a reality. Mary Lay, co-editor of *TCQ* from 1991 to 2003, has been equally generous in assisting with the current transition of the journal to Utah State. Mary and her staff's meticulous attention to detail has been crucially important in helping us sustain the day-to-day operations of *TCQ*.

Others, especially ATTW officers Jo Allen and Ann Blakeslee, have also been instrumental in assuring a smooth transition. Finally, we are deeply indebted to Carolyn Rude, who has undertaken the herculean task of guest editing two consecutive issues of *TCQ* in 2004, allowing us to get our bearings as we assume our journal responsibilities.

Our vision as editors is to preserve the distinguished legacy of *TCQ*, while at the same time introducing new features and initiatives. For example, we plan to continue *TCQ*'s tradition of sponsoring special issues that deal with significant topics in the field, because we believe these special issues are among the journal's most noteworthy and valuable features. The extensive examinations of topics provide touchstones for research in the field, serving the needs of established scholars working in specific areas, as well as doctoral students who are accumulating knowledge about a subject in anticipation of their own investigations or exams. A secondary, but equally important, function of these special issues is that they provide opportunities for more widespread and different types of participation in the journal, involving the ATTW community at-large in the organization's flagship research publication. During 2004, readers can anticipate more than the typical number of special issues as the regular editorial responsibilities for the journal shift from one institution to another.

A new feature in the journal is an annual interview with a major researcher or theorist who does not typically publish in technical communication journals but whose work is clearly influencing research in our field. The first such interview is scheduled to appear in *TCQ* during the next year.

In addition to introducing these features, we would like to initiate an expansion of the number of sources that index *TCQ*. We want to see the journal indexed in such places as the ISI Web of Knowledge, which includes online searching of the Social Sciences Citation Index, and the Arts & Humanities Citation Index. It is worth noting that most other journals where people associated with the field routinely publish (e.g., *JBTC, Technical Communication, Written Communication,* and *IEEE Transactions on Professional Communication*) are all indexed in the ISI Web, but *TCQ* is not. We believe that expanding the number of research venues in which *TCQ* scholarship is available will have the net effect of increasing awareness about the scholarly work we do.

Finally, as the new editors of *TCQ*, we want to increase the visibility of the journal at national conferences and foster open dialogue about it. Toward that end, we plan to schedule a session devoted to a discussion about the journal at every annual ATTW Conference. Such sessions will provide the members of ATTW with an opportunity to talk with the editors about their ideas for *TCQ* and to offer ideas for articles and special issues that would have broad organizational appeal.

While remaining committed to the high standards of scholarship that already characterize *TCQ*, we want to encourage research that looks forward to the possibilities of what technical communication practice and pedagogy may become. As participants in and observers of the field, we all already know that communicative practices across all social organizations are changing rapidly. To keep pace with these changes and support the field's understanding of what they mean, research must always be extending in new directions. Our plan is for the journal to support this important work.

The most obvious changes to the communicative practices that we study are associated with the introduction and evolution of new communication technologies, but that it not the whole story. The larger story is a complex one in which internationalization, legislation, and changing social paradigms also all play a role in how communicative practices come to be. It is our position that technical communication is not merely a field being carried along in this tide of history; it also represents a shaping force in the unfolding story.

For this reason, we would like to encourage scholarship in *TCQ* that reflects the trajectory of this story. A part of that story involves what takes place in the classroom. Another part includes what is happening in workplaces. Still another part involves what is happening in larger and often less well-defined social structures. The potential plot twists and dramatic turns in this story are not known, but it is clearly one in which technical communication plays a key role. Therefore, where it is productive, we want to encourage research that extends knowledge that is already part of the conversation in our field. At the same time, we welcome research that moves into new areas—whether that means reporting emerging communicative practices that may have not been explored in the field before or introducing new theoretical perspectives that yield telling views of the field.

We look forward to serving as *TCQ*'s editors.

Charie Thralls and Mark Zachry

Results of a Survey
of ATTW Members, 2003

David Dayton
Southern Polytechnic State University

Stephen A. Bernhardt
University of Delaware

This article presents the results of an April 2003 electronic survey of ATTW members. Results and interpretations are categorized as follows: a professional profile of respondents; member observations about ATTW and its activities (member participation, appraisal of benefits, and preferred topics for *TCQ*); and current issues and views of the field's future.

INTRODUCTION

In planning this special issue of *Technical Communication Quarterly* on the state of our profession, Carolyn Rude invited us to survey ATTW members to determine who they are, what they value about their membership, how much they participate in the organization's activities, what issues most concern them, and their hopes for the future growth of the field. We report the results of that survey here, with brief comments at the end reflecting our individual perspectives on the findings.

The only previous survey of ATTW members was conducted in 1993 by Jo Allen, with Steve Harding and Kristin White. Results were published in the Fall 1994 issue of the *ATTW Bulletin*. That survey was more limited in scope than ours, but it gathered data on gender and salaries that we did not include. (See the report of the 2003 STC salary survey in the article by Sandi Harner in this issue.) Where we can compare our data to theirs, we do so in the following.

METHODOLOGY

We administered the survey electronically in April 2003. We contacted members using a list of e-mail addresses provided by the ATTW Administrative Assistant, with the blessing of the Executive Committee. The initial list contained 422 e-mail addresses, 86% of the list of 488 individual members. We discovered that 26 e-mail addresses were invalid, and we managed to correct 14 of them. Thus, our e-mailed invitations to take the survey presumably reached 410 individual ATTW members, or 84%.

We sent out a total of three e-mail messages to members over the span of a week (April 23 to 30, 2003). All three messages contained a link that, when clicked, took respondents to an informed-consent form. Clicking on the "I agree" link at the bottom of the form brought up the survey itself, which consisted of forty-nine numbered items organized into six sections on a single HTML page. We estimated it would take respondents about twenty minutes to fill out the form. (The survey form is available through 2004 on the ATTW Web site under "ATTW Information.")

In all, we collected data from 228 returned forms, 56% of the ATTW members we presumably contacted (410), and 47% of the individual members (488). With that population size and a confidence level of 95%, our sample of 228 provides a maximum confidence interval of plus or minus 4.7 percentage points. We believe the results of our survey are fairly close to the results we would have obtained had we been able to gather data from all 488 individual members of ATTW. However, we can only be moderately confident about this, because we did not gather data from 53% of the individuals who were ATTW members at the time of the survey.

Most of the members we did not contact had not provided an e-mail address to ATTW. In some cases, that fact might be associated with attitudes that would tend to make their answers to the survey, as a group, significantly different from the answers provided by our respondents. We also assume that the electronic means of administering the survey may have been perceived negatively by some whom we did succeed in contacting but who elected not to complete the survey. Attitudes underlying the nonresponsiveness of some who received our e-mail might have led them, if they had taken the survey, to submit data that would have significantly altered the proportions and patterns of opinions we summarize in this article.

Data submitted from the survey form went into one of our e-mail accounts (ddayton@spsu.edu). Dayton used Survey Solutions for the Web (SSW, http://www.perseusdevelopment.com) to read the e-mail containing the data directly into a database application within the survey software program. Dayton used SSW to create a Word document containing data tables and lists of open-ended responses for all the survey items. He also exported the data to Excel to facilitate data inspection.

Many respondents used the open-ended text-entry boxes on the survey (the "other" option) that accompanied the lists of fixed-choice answers. Examining these "other" explanations in Excel, Dayton was able to sort some of them into fixed-choice categories by logical imputation. Similarly, inspecting the data closely, Dayton could resolve contradictions and fill in blanks in the answers for particular records. For instance, when a respondent indicated that he or she attended the ATTW conference in Chicago in 2002 but failed to indicate any participation in an ATTW conference during the past two years, even as attendee, Dayton imputed an oversight to that respondent and counted the respondent as an attendee in tallying responses for the item on conference participation. The changes to the data resulting from this process were minor; they decreased the number of responses counted as "other" or "unanswered" for some items and reduced discrepancies between respondents' answers to items eliciting the same or overlapping information.

In this article, the data tables for fixed-choice items summarize the data as edited by Dayton. The complete, original data tables are available in the members-only section of the ATTW Web site, along with a read-only Excel file containing the unedited data for fixed-choice items. Open-ended responses not associated with an "other" fixed-choice answer were removed from the Excel file to obviate any concern that a respondent could be identified by examining individual data records. The complete collection of responses to the open-ended items, however, is available in a Word file that is also available in the same area of the ATTW Web site.

One author (Bernhardt) sorted the open-ended responses into emergent categories, aiming for five to eight categories of response for each question. Similar responses were grouped, categories with only a few responses were consolidated with the most similar category, and frequently mentioned issues were identified for inclusion in the results tables presented at the end of the results section.

RESULTS

We have organized our presentation of the survey results under the following three heads:

- A Professional Profile of Respondents
- Participation, Appraisal of Benefits, and Topics for *TCQ*
- Current Issues and Views of the Field's Future

We present the results mainly in tables introduced by brief summaries highlighting the most noteworthy data.

A Professional Profile of Respondents

In this section, we review the data about respondents' current employment, degrees they held, and the types of programs in which they taught. (We use the past tense in reporting data, recognizing that this report is a cross-sectional snapshot of a group.)

Nine of ten were college or university teachers.   Nine out of ten respondents taught in institutions of higher education (see Table 1). Three-fourths were faculty members at a college or university. Ten teaching respondents were adjuncts, 16 taught at two-year colleges, and 21 were graduate students. The remaining 7% of respondents were divided among practitioners (10), retirees (3), and university administrators (2).

Of the 181 teaching respondents (out of 206 total) providing information about full-time versus part-time status, 152 (84%) were employed in full-time positions. Three-fourths of the part-timers were graduate teaching assistants. Of the 179 teaching respondents providing information about tenure status, 78 (44%) already had tenure, and another 57 (32%) reported that they held tenure-track positions. About one in four (44) held nontenure track positions.

As Table 2 shows, assistant professors constituted the largest group of teaching respondents holding faculty or adjunct positions—38%. Professors were the next largest group (23%), followed by associate professors (21%).

The vast majority held doctorates.   Seven of 10 respondents held doctorates (see Table 3), and another one in four held master's degrees. Of the 64 respondents who did not have a doctorate, 37 (or 58%) were enrolled in doctoral programs. The respondents were about evenly split, 49% to 51%, between those awarded their highest degree in the past six years and those awarded their highest

TABLE 1
Employment Categories of Respondents

Employment Category	Count	% of 228 Respondents
College or university faculty	175	77%
College or university adjunct	10	4%
GTA (21), GRA (1), ABD (2), graduate student (4)	28	12%
Professional practice of technical communication	10	4%
Retired	3	1%
University/college administrator	2	1%

TABLE 2
Job Titles Reported by Nongraduate Student Teaching Respondents

Job Title	Count	% of 185 Respondents
Professor	42	23%
Associate professor	38	21%
Assistant professor	71	38%
Instructor	16	9%
Lecturer (11), sr. lecturer (2), college lecturer (1)	14	8%
Other (2 no answer, 1 writing coach, 1 no rank)	4	2%

TABLE 3
Highest Academic Degree

Academic Degree	Count	% of 228 Respondents
PhD	157	69%
Doctor of arts	7	3%
Master's	57	25%
Other (4 bachelor's, 1 associate's, 2 ABD)	7	3%
Total	228	100%
Number now enrolled in PhD program	37	16%

degree more than six years ago (see Table 4). A substantial portion of respondents (38%) had been in the field for more than ten years.

Doctorates were mainly in three areas. The survey asked respondents to categorize the area in which their highest academic degree had been awarded. Three distinct disciplinary rubrics accounted for 84% of the doctorates. As Table 5 shows, English studies and literature accounted for 31%, and technical/professional communication, with and without "and/or rhetoric," accounted for 30%. Another 23% of respondents identified composition studies and/or rhetoric/composition or rhetoric alone as the area in which they had earned their highest degree. Moreover, of the 62 respondents who had been members for more than ten years, a

TABLE 4
Years Since Highest Degree Conferred

Years	Count	% of 228 Respondents
1 to 3 years	63	28%
4 to 6 years	49	21%
7 to 10 years	29	13%
More than 10 years	87	38%

TABLE 5
Area of Respondents' Highest Academic Degree

Discipline	Count	% of 228 Respondents
English studies (63) and literature (7)	70	31%
Technical/professional communication and/or rhetoric	52	23%
Composition studies or rhetoric/composition	34	15%
Rhetoric	19	8%
Technical/professional communication	16	7%
English education (8) and linguistics (5)	13	6%
Other	24	11%

solid majority (34, or 55%) indicated that their highest degree was in English studies or literature. The meaning seems clear: Although more respondents held their highest degree in English studies and literature than in any other area, this doctoral specialization is destined to fade as those associated with technical and/or professional communication gain ascendancy.

The survey also asked respondents enrolled in doctoral programs to categorize their program using the same disciplinary labels shown in Table 5. The distribution of responses on this item shifted dramatically in favor of technical/professional communication: 19 of the 37 doctoral students, or 51%, were in programs labeled as technical/professional communication, with and without "and/or rhetoric" (see Table 6).

Programs in technical/professional communication predominated. Three items on the survey asked respondents to categorize the undergraduate, master's, and doctoral programs in which they were teaching. At all three levels, programs specializing in technical/professional communication were the most numerous, but the percentage was highest at the undergraduate level and the smallest at the doctoral level (see Tables 7, 8, and 9). At all three levels, the second most common

TABLE 6
PhD Programs of Doctoral-Student Respondents

Disciplinary Rubric of Doctoral Program	Count	% of 37 Respondents
Technical/professional communication and/or rhetoric	16	43%
Composition studies or rhetoric/composition	5	14%
Rhetoric	5	14%
Technical/professional communication	3	8%
English education (1) and linguistics (1)	2	5%
English studies	1	3%
Other: 3 no answer, 1 ed., 1 texts & technology	5	14%

TABLE 7
Types of Undergraduate Programs in Which Respondents Teach

Type	Count	% of 228 Respondents
Program has a specialization or major for technical communicators	102	45%
English studies program with some technical/professional communication courses	42	18%
Service courses only	24	11%
Program has a minor in technical/professional communication	23	10%
Only a certificate program in technical/professional communication	11	5%
No answer provided	26	11%

TABLE 8
Types of Master's Programs in Which Respondents Teach

Type	Count	% of 228 Respondents
Program dedicated to technical/professional communication	69	30%
English studies or rhetoric/composition program with some technical/professional communication courses	33	14%
English studies program with concentration in technical/professional communication	27	12%
Other master's program	13	6%
Certificate program only in technical/professional communication	3	1%
Do not have a master's program	56	25%
No answer provided	27	12%

TABLE 9
Types of Doctoral Programs in Which Respondents Teach

Type	Count	% of 228 Respondents
Program dedicated to technical/professional communication	37	16%
English studies doctoral program with concentration in technical/professional communication and/or rhetoric	20	9%
Other doctoral program unrelated to technical/professional communication	14	6%
English studies doctoral program with some technical/professional communication courses	11	5%
Do not have a doctoral program	118	52%
No answer provided	28	12%

type of program was an English studies and/or composition/rhetoric program offering some courses or a concentration in technical/professional communication.

Affiliations with other professional organizations.   The survey included two items requesting information about membership and participation in other professional organizations. The data collected from those items are summarized in Tables 10 and 11. The two professional organizations with the most membership overlap among ATTW respondents are the Conference on College Composition and Communication (CCCC) and Society for Technical Communication (STC). Almost two-thirds of the respondents (145) belonged to the CCCC of the National Council of Teachers of English (NCTE)—only slightly more than belonged to the

TABLE 10
Memberships in Other Professional Organizations

Professional Organizations	Count	% of 228 Respondents
NCTE/CCCC: National Council of Teachers of English, Conference on College Composition and Communication	145	64%
STC: Society for Technical Communication	135	59%
CPTSC: Council for Programs in Technical and Scientific Communication	73	32%
MLA: Modern Language Association	63	28%
ABC: Association for Business Communication	39	17%
IEEE Professional Communication Society	35	15%
TYCA: Two-Year College English Association	16	7%
ACM-SIGDOC Association for Computing Machinery	14	6%
NCA: National Communication Association	10	4%
Other(s)	64	28%

TABLE 11
Annual Conferences Attended at Least Once in 2001 and 2002

Annual Conference of Professional Organization	Count	% of 228 Respondents
CCCC: Conference on College Composition and Communication	114	50%
ATTW: Association of Teachers of Technical Writing	92	40%
CPTSC: Council for Programs in Technical and Scientific Communication	52	23%
MLA: Modern Language Association	45	20%
STC: Society for Technical Communication	45	20%
ABC: Association for Business Communication	23	10%
Computers and Writing	23	10%
RSA: Rhetoric Society of America	20	9%
IPCC: International Professional Communication Conference (IEEE PCS)	18	8%
SIGDOC: Documentation Special Interest Group of Association for Computing Machinery	11	5%
NCA: National Communication Association	10	4%
NCTE: National Council of Teachers of English	10	4%
CHI: Computer-Human Interaction special interest group of Association for Computing Machinery	2	1%
UPA: Usability Professionals Association	3	1%
Other(s)	45	20%

STC (135). Of the 145 ATTW members who reported CCCC membership, 81 also belonged to STC, 54 to the Council for Programs on Technical and Scientific Communication (CPTSC), and 47 to the Modern Language Association (MLA). When the 135 ATTW/STC members are used as the basis of comparison, the only notable change in the numbers occurs with MLA members: 38 also belong to MLA.

The CCCC annual conference was attended by more respondents than any other in 2001 and 2002, followed by the ATTW conference (which, of course, is held in conjunction with the annual conference of the CCCC). Substantially fewer respondents attended the conferences held by CPTSC, STC, and MLA in 2001 and 2002 (see Table 11).

Participation, Appraisal of Benefits, and Topics for *TCQ*

In this section, we review the data about respondents' years of membership in ATTW, their level of participation in the organization and its primary activities, their appraisal of membership benefits, and their level of interest in a number of topics suggested by the survey as topics *TCQ* could prioritize for future issues.

Newer members more numerous, much less active on committees.   The distribution of respondents by the number of years they have been ATTW members shows a healthy spread, somewhat weighted in favor of new and recent members (see Table 12). Almost 40% had been members for one to three years, about one-third had been members for four to six years, and about one-fourth had been members for ten years or longer. Table 12 presents this frequency distribution along with the data showing the number and percentage of respondents indicating active or limited participation in one or more ATTW committees during the two years prior to the survey.

Overall, only about one in five respondents reported active or limited participation on one or more ATTW committees during the two years prior to the survey. In the one to three years of membership category, only 4 of 89 members (5%) reported some committee participation. Much higher proportions of respondents who had been members four to ten years (26%) and over ten years (35%) reported participating on one or more committees.

Strong attendance at ATTW conference, weak at MLA sessions.   Over half the respondents indicated that they attended the ATTW annual conference in 2003 or 2002, and about one in four presented a paper at one or both conferences

TABLE 12
Years of ATTW Membership and Committee Participation

Years ATTW Member	Count	% of 227 Answers	Active or Limited Committee Participation	% of Years ATTW Member Category	Percent of 46 on Committees
1–3 years	89	39%	4	5%	9%
4–10 years	76	34%	20	26%	43%
> 10 years	62	27%	22	35%	48%
Total	227	100%	46	—	100%

(see Table 13). The ATTW-sponsored sessions at the annual conference of the MLA drew far less participation, with only 13% of respondents indicating they attended the sessions during 2002 and 2001.

Over half read ATTW-L, a third do not.   Almost two-thirds of respondents reported that they read the ATTW-L listserv messages always (38%) or sometimes (27%), but about one-third indicated they never read them (see Table 14). As with most listservs, the number of relatively frequent posters was a small fraction of the

TABLE 13
Participation in ATTW Conference and MLA Sessions

Conference Participation in Past Two Years	Count	% of 228 Respondents
ATTW Conference		
Attendee	123	54%
No participation in past two years	97	43%
Presenter	61	27%
Panel chair	22	10%
Organizer/staff	16	7%
Item not answered	8	4%
MLA-ATTW Sessions		
No participation in past two years	178	78%
Attendee	29	13%
Presenter	11	5%
Panel chair	2	1%
Organizer/staff	5	2%
Item not answered	21	9%

TABLE 14
Participation in ATTW-L Listserv

ATTW-L Listserv Participation in Past Two Years	Count	% of 228 Respondents
Reader		
Always	86	38%
Sometimes	61	27%
Never	74	32%
Item not answered	7	3%
Contributor		
No contributions in past two years	127	56%
Occasionally	55	24%
Two to six times a year	27	12%
More than six times a year	9	4%
Item not answered	10	4%

number of readers. Only 9 respondents (4%) indicated that they had posted to the listserv six or more times a year during the past two years. Three times that many (27, or 12%) said they posted listserv messages two to six times a year, while 55 (24%) indicated that they had posted occasionally.

ATTW Web site.   Nine out of 10 respondents indicated that they had used at least one of the Web site services listed in Table 15 during the past two years. The most frequently mentioned services were information about the ATTW conference, *TCQ*, calls for papers, and teaching resources, along with specific administrative information about the organization.

Conference attendance and views on linkage with CCCC.   Among the 228 survey respondents, attendance at the annual ATTW conference during the past six years ranged from a low of 19% in 1999 to a high of 37% in 2003 (see Table 16). On another item, about two-thirds of respondents who answered (133 of 195) indicated that they appreciated the linkage of the ATTW conference with the CCCC convention, because they usually attended CCCC anyway. About one-third said they would prefer an independent conference, most of them specifying that they would like such a conference to be held in the fall.

Asked to rate the importance of ATTW membership benefits on a four-point sale, almost nine of ten respondents rated their subscription to TCQ "very impor-

TABLE 15
Web Site Services Used in Past Two Years

Web Site Services Used in Past Two Years	Count	% of 228 Respondents
Number indicating usage of at least one web site service	206	90%
ATTW conference information	126	55%
Information about *Technical Communication Quarterly (TCQ)*	113	50%
Calls for papers	110	48%
General ATTW information (officers, committees, constitution, etc.)	109	48%
Teaching resources	106	46%
Online (PDF) copies of selected *TCQ* articles *(members only area)*	95	42%
Academic programs information	90	39%
ATTW news	90	39%
Membership directory *(members only area)*	88	39%
Job announcements	84	37%
Online (PDF) bibliography *(members only area)*	79	35%
Professional resources (proposal writing, bibliographies, organizations, journals)	68	30%
Other ATTW publications information	67	29%
Online (PDF) bulletin *(members only area)*	55	24%

TABLE 16
Respondents' Attendance at Last Six ATTW Conferences

Conference Year and Location	Count	% of 228 Respondents
1998 Chicago	49	21%
1999 Atlanta	44	19%
2000 Minneapolis	56	25%
2001 Denver	60	26%
2002 Chicago	82	36%
2003 New York	84	37%

tant." We calculated an index score for each benefit, with "very importnat" worth 3 points, "somewhat important" worth 2 points and "not very important" worth 1 point. Table 17 shows the rank order of benefits based on this index score, along with the number rating each benefit "very important" and the corresponding percentage of the total number of respondents. The resulting order coincides with an ordering based on the number rating each benefit "very important."

Much higher usage of TCQ than most other journals. Not only was a subscription to *TCQ* the number one membership benefit, according to respondents, it was also the journal they read more than any of the nearly two dozen others we asked about. Sixty percent of the respondents said they used the bibliography published in the fall issue of *TCQ* each year (see Tables 18, 19). Almost half reported that they read or skimmed nine or more of the sixteen articles that *TCQ* published in 2002. Another 30% said they read or skimmed five to eight articles. Contrasting with the high rate of readership among members was the low rate of article submissions to *TCQ*: Seventeen respondents reported submitting an article in 2002, and only two reported that they submitted book reviews.

TABLE 17
Ratings of ATTW Membership Benefits

Benefit	Number Answering Very Important	% of 228 Respondents	Index Score
Subscription to *Technical Communication Quarterly (TCQ)*	201	88%	651
Feeling of connection to a professional community	163	71%	600
Free use of *TCQ* articles for coursepacks	140	61%	543
Member-restricted areas of ATTW web site	129	57%	511
Annual conference participation	105	46%	491
Member directory	68	30%	444
Bulletin (newsletter)	49	21%	406

TABLE 18
Respondents' Usage of *Technical Communication Quarterly*

Usage	Count	% of 228 Respondents
Use the bibliography published in the fall issue	137	60%
Read/skimmed 13 to 16 articles published in 2002	54	24%
Read/skimmed 9 to 12 articles published in 2002	47	21%
Read/skimmed 5 to 8 articles published in 2002	68	30%
Read/skimmed 1 to 4 articles published in 2002	49	21%
Read/skimmed 0 articles published in 2002	4	2%
Submitted an article in 2002	17	7%
Submitted a book review in 2002	2	1%

TABLE 19
Reading of Journals Other than *TCQ*

Journal Title	Number Answering Most or Several Articles Each Year	% of 228 Respondents	Index Score[a]
Technical Communication	161	71%	435
College Composition and Communication	121	53%	332
Journal of Business and Technical Communication	117	51%	318
College English	82	36%	263
Journal of Technical Writing and Communication	77	34%	247
IEEE Transactions on Professional Communication	72	32%	242
Journal of Business Communication	59	26%	206
Business Communication Quarterly	55	24%	187
Written Communication	42	18%	173
JAC: A Journal of Composition Theory	38	17%	169
Computers and Composition	40	18%	156
Rhetoric Review	37	16%	138
Rhetoric Society Quarterly	34	15%	125
Kairos	34	15%	118
Writing Program Administration	28	12%	104
Journal of Computer Documentation	14	6%	67

[a]Valuation for index score: most articles = 3, several articles a year = 2, rarely/for particular articles = 1, never = 0. Score obtained by multiplying the valuation of the interest category by the number of respondents rating the topic in that category and then adding up the three subtotals.

On a separate item of the survey, we listed twenty-three professional and scholarly journals besides *TCQ* and asked respondents to indicate whether they read most articles, several articles a year, rarely/for particular articles, or never. Using a method similar to the preceding one, we created an index score to summarize the relative rank of journals based on the reading ratings given by respondents.

- The index score for *Technical Communication* was substantially higher than for any other journal, as shown in Table 19. Seven of 10 reported reading at least several articles a year in *Technical Communication*, the journal of the Society for Technical Communication.
- *College Composition and Communication* and the *Journal of Business and Technical Communication* had similar index scores, and about half the respondents said they read at least several articles a year in those journals.
- The third tier of the journal popularity hierarchy that emerges from the data in Table 19 consists of *College English*, the *Journal of Technical Writing and Communication*, and *IEEE Transactions on Professional Communication*.
- The *Journal of Business Communication* (*JBC*) and *Business Communication Quarterly* (*BCQ*) constitute a fourth tier of popularity, with similar numbers. Over three-fourths of the 59 respondents who read at least several articles a year in *JBC* also read at least the same amount in *BCQ*.
- A larger fifth tier of popularity comprises journals focusing on rhetoric, writing theory and research, writing program administration, and technology and writing: *Written Communication, JAC: A Journal of Composition Theory, Computers and Composition, Rhetoric Review, Rhetoric Society Quarterly, Kairos,* and *Writing Program Administration.*

Interest ratings for possible TCQ topics: A familiar menu. We tested the patience of survey takers (☺) by asking them to indicate their level of interest in a plethora of topics for articles in future issues of *TCQ*. We listed fifty-eight topics organized into four general categories: Teaching; Workplace Practice; Theory, Methods, the Discipline; and Specialized Discourses. To condense the results, we calculated an index score for each topic based on the interest ratings it received (see the note under Tables 20 and 21). The average index score for all topics was 470. Topics receiving a higher than average index score are listed in Table 20; those receiving a lower score are shown in Table 21.

Table 20 can be parsed in any number of ways, but we found it interesting to note these groupings of topics scoring over 500 on the interest index:

- Visual communication, applications of rhetorical theory to technical communication, and rhetoric of technology produced an average index score of 556.
- Roles of technical communicators and intersections with other fields, although listed in distinct topical categories, seem related in their boundary-spanning implications. Together, they produced an average index score of 554.

TABLE 20
Topics Receiving Above Average Interest (Index ≥ 470)[a]

Topic	Index Score	Topic	Index Score
Teaching		Theory, Methods, the Discipline	
Visual communication	561	Intersections with other fields	562
Teaching with technology	546	Applications of rhetorical theory to TCOM	557
Research methods	517	Ethics	525
Online teaching	504	Contextual inquiry and workplace ethnography	518
Undergraduate curriculum: Majors	491	Cultural studies of technical writing	490
Digital media	494	Human factors theory/research	489
Graduate curriculum: MA/MS	470	Language and learning theory	472
Distance learning	470	Specialized Discourses	
Workplace Practice		Rhetoric of technology	550
Usability	547	Hypertext and digital rhetorics	491
Roles of technical communicators	546	Civic rhetoric and public policy discourse	484
Human-centered design	541	International communication	478
Information architecture	540	Rhetoric of science	474
Audience analysis	535		
Knowledge management	521		
Collaboration	502		
Content management	498		
Project management	485		
Documentation	476		
Editing	470		

[a]Valuation for index score: high interest = 3, some interest = 2, little interest = 1, no interest = 0. Score obtained by multiplying the valuation of the interest category by the number of respondents rating the topic in that category and then adding up the three subtotals.

- Usability, human-centered design, and information architecture garnered three of the top four scores in the Workplace Practice category, producing an average index score of 543.
- Two topics that are foundational to the concerns of our discourse community, audience analysis and ethics, received similarly high interest ratings, producing an average index score of 530.
- The topic listed simply as "research methods" together with one particular method—contextual inquiry and workplace ethnography—had nearly identical index scores of 517 and 518, respectively.

Also scoring over 500 were teaching with technology (546), online teaching (504), knowledge management (521), and collaboration (502).

It is interesting to compare the list of topics just summarized with the areas of interest receiving the highest interest ratings on an ATTW survey of members car-

TABLE 21
Topics Receiving Below Average Interest (Index < 470)[a]

Topic	Index Score	Topic	Index Score
Teaching		Theory, Methods, the Discipline	
Industry training	461	Situated learning	451
Writing style	458	History of technical communication	450
Assessment	456	Survey research methods	450
Genres	453	Genre theory	436
Project management	449	Gender studies	396
Science writing	448	Activity system theory	384
Graduate curriculum: PhD	437	Specialized Discourses	
Service courses	430	Environmental writing	451
Internships	426	Health and medical communication	449
Service learning	424	History of rhetoric	430
Engineering writing	423	Legal and regulatory writing	418
Teaching in two-year colleges	212	Workplace Practice	
		Technology innovation management	428
		Policies and procedures	418
		Single-sourcing methods/technologies	386
		XML, implementation, and impacts	378

[a]Valuation for index score: high interest = 3, some interest = 2, little interest = 1, no interest = 0. Index score obtained by multiplying number of respondents rating the topic in each interest category and then adding up the three subtotals.

ried out by Jo Allen et al. in 1993 and published in the Fall 1994 *ATTW Bulletin*. (A PDF copy of the pages summarizing the results of that survey can be found on the ATTW Web site, along with other materials related to this article.) In that survey ten years ago, ATTW members rated the following topics of highest interest, in descending order: pedagogy, writing process, document design, style, and editing. The only two that we thought to include in our survey, style and editing, produced lower than average interest ratings. And, of course, we listed many topics that were not even in the vocabulary of the field ten years ago.

Current Issues and Views of the Field's Future

In the final section of our survey, we asked respondents to select from a list of names we use for our academic programs the one they preferred. In a follow-up item, we asked them to share their thoughts on the naming issue. Finally, we presented them with five open-ended questions aimed at gathering their views on current issues and the directions they would like to see the field take going forward.

What to call the field.   The preceding data reviewed on the disciplinary rubrics for academic degrees and programs with which the survey's respondents were affiliated (see Tables 5 through 9) amply demonstrate what is common knowledge among us: We are a field that goes by several names. What to call the field is a question that will not soon be settled, if it ever is. Nevertheless, we thought it would be interesting to take a poll of ATTW members to see which names are preferred. We presented a list of names, added a specified "other" choice, and asked respondents to vote for only one. The results are shown in Table 22.

"Technical Communication" garnered the most votes, 88, 16 more than the runner-up, "Professional Communication" (39% and 32%, respectively). However, the combined categories "Professional Communication/Writing" and "Technical Communication/Writing" get an equal number of votes, 95 each, or 42%. The remaining 16% of respondents divided their loyalties among "Rhetoric," "Writing Studies," "Texts and Technology," and "Digital Rhetorics," in that order of frequency, but "no answer" was the second most popular choice of this minority grouping.

In the open-ended follow-up to the fixed-choice poll on the best name for the field, respondents suggested a variety of other names, including these:

- "Workplace Communication" (2) and "Workplace Writing" (1)
- "Applied Rhetoric" (3) and "Applied Professional Rhetoric" (1)
- "Information Design" (2)

Some 21 respondents noted the obvious: It is difficult to find right terms for naming the field. Three argued that the name should communicate plainly to outsiders, but 8 said the name should vary with the context. One wrote, perhaps facetiously, "I want a name that includes all areas (technical, scientific, medical, business, etc.), that includes all types of communication (writing, print and online,

TABLE 22
Preferences for What to Call the Field

Preferred Name for the Field	Count, Fixed Choice	% of 228 Respondents
Technical Communication	88	39%
Professional Communication	72	32%
Professional Writing	23	10%
Rhetoric	14	6%
Writing Studies	9	4%
Technical Writing	7	3%
Texts and Technology	5	2%
Digital Rhetorics	2	1%
No answer provided	8	4%

multimedia, etc.), that includes rhetoric and probably linguistics and discourse studies." Others just did not care: "I don't have a problem with the diversity with which we name ourselves. I, myself, use the terms *professional communication, writing studies,* and *rhetoric and communication,* depending on the context and audience I'm speaking with." And at least one did not accept the claim in the fixed-choice question that "this field has struggled with naming itself," writing, "I have been a Technical Communicator for more than twenty years and don't recall any struggle over the name."

We invite readers to peruse the many views on this and other issues raised by the open-ended questions. The ATTW Web site has a Word document with all the responses. We believe the following comment deserves to be the last word here on the naming issue:

> Naming the field is a consequence of the field's lack of a coherent identity and lack of professional status. It's probably also a consequence of our historical role as a practice that is inevitably subordinate to other professional practices and to management structures outside of our own field. Thus, the organization makes up its own name for positions held by technical communicators, and we tend to think that's great as long as our identity gets linked with other higher-status professions such as engineers or managers, as in "knowledge engineer" or "information manager." Finally, it's probably a consequence of the rapid changes in the work people are doing and the new areas in which we practice. Names can't keep up with the changing and expanding roles we fill.

The most important skills students need to succeed as professionals.   We asked respondents what they considered to be the most important skills for their students' success as technical communicators. Bernhardt analyzed the brief answers submitted by 178 respondents (as he did all of the data for the open-ended questions). He found that two related categories predominated and were equally balanced: rhetorical skills, and writing and editing skills. Of the other categories of skills, there is a balanced representation down the list shown in Table 1. (For all open-ended questions, some respondents had multiple responses. All counts represent the number and percentage of respondents who mentioned a topic or issue out of the total number of respondents, not the percentage of a particular response out of all responses.)

The groupings that define skill areas in Table 23 are somewhat arbitrary. Document design might be considered part of general writing and editing skills, just as research skills might be viewed as a form of specialized expertise. Collaboration and teamwork could obviously be combined with oral and interpersonal skills. Table 24, then, offers a less granular representation than Table 23 of the opinions about the most important skills students need to succeed in the workplace.

TABLE 23
Most Important Skills for Students to Succeed as Professionals: More Granular

Skill Area	n (%) of 178	Frequent Descriptors
Rhetoric	114 (64%)	Audience analysis, ability to adapt communication to situation, genre knowledge, understanding of rhetorical situation, rhetorical problem solving
Writing and editing	112 (63%)	Style, correctness, organization
Technology	58 (33%)	Facility with, critical understanding of, and ability to learn technologies
Personal traits and work skills	54 (30%)	Flexibility, ethics, organization, humor, ability to learn, professionalism, attention to detail, time management, cultural awareness
Specialized expertise	48 (27%)	Project management, business practices, scientific and technical knowledge
Document design	43 (24%)	Visual communication, format, graphics, usability, user-centered design
Problem solving/thinking skills	36 (20%)	Creative problem solving, analysis, critical thinking, problem solving
Collaboration and teamwork	35 (20%)	
Oral or interpersonal communication	28 (16%)	Presentation skills, interpersonal, interviews, listening
Research	14 (8%)	Ability to do research; familiarity with research literature

TABLE 24
Most Important Skills for Students to Succeed as Professionals: Less Granular

Skill Area	n (%) of 178	Frequent Descriptors
Writing, editing, and document design	155 (87%)	Style, correctness, organization
Rhetoric	114 (64%)	Audience analysis, ability to adapt communication to situation, genre knowledge, understanding of rhetorical situation, rhetorical problem solving
Personal traits, work skills, problem solving	90 (51%)	Critical thinking, analysis, flexibility, ethics, organization, humor, ability to learn, professionalism, attention to detail, time management, cultural awareness
Communication	63 (35%)	Collaboration, teamwork, presentation skills, interpersonal, interviews, listening
Specialized expertise	62 (35%)	Research skills, project management, business practices, scientific and technical knowledge
Technology	58 (33%)	Facility with, critical understanding of, and ability to learn technologies

No matter how we look at the responses, reviewing the lists of important skills suggests a useful way for programs to identify both for themselves and their students the intended outcomes of instruction and the importance of a broad and useful skill set for success in technical communication.

A sampling of quotes.   Most important is...

The ability to understand how texts function (intentionally and unintentionally), to assess audience needs, to analyze their own positions in organizations, to make appropriate rhetorical choices.

Overall, a sense of themselves as citizens within a larger community, which in my mind includes a sense of one's ethical practices. Then audience analysis skills, document design skills, and technology skills.

Reflectiveness—the ability not only to apply effective rhetorical techniques, but to understand how those techniques work, how to explain their merits to others, and how to refine or adapt them for future occasions. Also important is a historical perspective, or at least a habit of mind that considers current practices and realities in light of other possibilities.

To be able to cut through the information overload, shape the appropriate information to the particular audience, and communicate the results with effective writing, formatting, and graphic selection for paper and Web documents

How are we doing at teaching the most important skills?   Overall, respondents expressed some confidence that their programs were teaching the important skills well (see Table 25). Over one-fourth of the respondents indicated that their programs were teaching all identified important skills well, offset by a few (8) who felt that no skills were being taught well, signaling dismay with their program, department, or resources. (Negative responses are summarized in Table 26.) Rhetoric was mentioned frequently, both for being taught well and needing to be taught better, with responses focusing on audience awareness and adaptation (as something writers need to do), but also, in more general terms, on theoretical understanding of language and situation. (Note that the question was biased by asking about important skills, as opposed to important knowledge or important theory or some other descriptor.)

Technology skills emerged as a specific area where teaching could be stronger, although no one area was mentioned as a needed area of improvement by more than 13% of the respondents. Comments on technology were directed not only at software skills, but also at critical understanding of technology and the ability to learn what is needed and apply appropriate tools.

Writing skills were described either as ability to write clearly, correctly, and well, or as document design, with an emphasis on visual communication and us-

TABLE 25
Important Skills Being Well Taught

Skills We Teach Well	n (%) of 159	Frequent Descriptors
Rhetoric	24 (15%)	Audience analysis, ability to adapt communication to situation, genre knowledge, understanding of rhetorical situation, rhetorical problem solving
Writing and editing	21 (13%)	Style, correctness, organization
Technology	16 (10%)	Facility with, critical understanding, and ability to learn technologies
Document design	13 (8%)	Design, format, usability, user-centered design, graphics
Problem solving/thinking skills	10 (6%)	
Personal traits and work skills	7 (4%)	Flexibility, organization, and humor, ethics, lifelong learning, professionalism, diversity, time management
Collaboration and teamwork	5 (3%)	
Oral or interpersonal communication	5 (3%)	Interviews, presentation skills, listening
Research	3 (2%)	
Specialized expertise	1 (1%)	Project management
All	44 (28%)	The skills we identify are taught well, for the most part

ability. Under either definition, responses that indicated these skills were being taught well were balanced by those who felt a need to improve the teaching of writing, editing, and document design. Some respondents elaborated on the central or foundational importance of a strong set of writing and editing skills.

In all, 159 respondents offered opinions on this question, most of which were divided into multiple parts for sorting into the summary representation shown in Tables 25 and 26.

A sampling of quotes.

I honestly believe we are trying to address the importance of effective written communication in the workplace and to include all the various forms they might encounter. We are certainly training them how to approach written communication and provide them with strategies to tackle tasks.

We don't do much with interpersonal skills directly, but we have a lot of collaborative projects.

We should be teaching or fostering an attitude of lifelong learning—and a sense of what it means to be a professional in the field.

TABLE 26
Important Skills Not Being Taught Well

Skills We Are Not Teaching Well	n (%) of 159	Frequent Descriptors
Technology	20 (13%)	Technology, Web design, multimedia, software skills not taught because of lack of current tools, faculty resistance or lack of skill
Rhetoric	19 (12%)	Audience awareness, cultural interpretation, reflective practice, disciplinary history, multiple genres, and general, transferable skills in analyzing texts, communicating effectively
Writing and editing	15 (9%)	Style, correctness
Document design	14 (9%)	Design, format, usability, user-centered design, graphics
Specialized expertise	10 (6%)	Understanding business culture, project management, and knowledge management, and we need a way to help students to an understanding of organizations and management
Research	5 (3%)	
Personal traits and work skills	5 (3%)	Fexibility, organization, and humor, ethics, lifelong learning, professionalism, diversity, time management
Oral or interpersonal communication	4 (3%)	Interviews, presentation skills, speaking, listening
Problem solving/thinking skills	4 (3%)	Creative problem solving, critical thinking
Collaboration and teamwork	1 (1%)	
Not teaching anything well	8 (5%)	Little coherence in program or poor teaching because of lack of faculty, resources, goals, or support

We don't have an actual "writing better" course. We admit people who write pretty well and then teach them to do different kinds of things with that skill—but I'm not sure we actually enhance their basic writing ability.

We underplay the importance of general, transferable skills in analyzing texts, communicating effectively, and solving problems.

I don't think the program I'm currently in as a PhD student has paid much attention to any of these skills explicitly in coursework, and I think it should. Like other programs, ours seems to follow industry practice in most ways rather than setting up a solid theoretical framework. There is very little "critical thinking" applied to "current practice;" rather, the focus is on learning "current practice."

The most significant issues facing technical communication graduates. Almost half of the respondents commented on the current downturn in the job market, noting that students were having trouble finding jobs, were having to relocate, or were choosing to continue their education. A full 25% of respondents commented on the difficulty of asserting the value of technical communication in the workplace, emphasizing the need for graduates to understand and market their skills in a business culture. These respondents tended to mention the importance of forming a professional identity, with a sure sense of the value of technical communication in various workplaces.

In all, 145 respondents answered this question, most offering multipart answers. Table 27 presents our summary representation of this data.

A sampling of quotes. The most significant issue facing graduates is...

This area is heavily invested in telecommunications and many in our STC chapter have lost jobs. We need to help our graduates think more broadly and break into areas of interest outside of software documentation.

How to find a job that doesn't construe them as automatons, that recognizes them adequately as professionals, and that doesn't work them to death.

Understanding the breadth and depth of technical communication's significance in industrialized cultures. It isn't just writing manuals!

Technological change is the big long-term issue—the changing nature of texts and communication.

How to become active managers of their own professional destinies and not just passive receivers of tasks to complete.

They are unaware of how critically important their work is and how important it is that they have high standards for themselves and their work.

The most significant issue facing undergraduate programs. When asked to name the most significant issue facing the undergraduate program in which they

TABLE 27
Most Significant Issues Facing Graduates of Technical Communication Programs

Issues Facing Grads	n (%) of 145	Frequent Descriptors
Jobs	69 (48%)	Finding good ones, keeping jobs, being able to move
Identity and adaptation	37 (25%)	Being able to market their skills in a business/industry culture; forming identities, adapting to business culture and finding satisfaction
Skills	15 (10%)	Technology, writing, collaboration, business
Theory/research	11 (8%)	Having the right knowledge and staying current
Don't know	22 (15%)	

taught, respondents focused on issues of recruiting qualified students and faculty, often in the face of resistance or "benign neglect" from departments concerned with literature or other concerns. A sizable proportion of respondents (19%) provided a diverse list of weaknesses in their programs, from not having a coherent program to difficulty maintaining sufficient course sections, to problems associated with keeping up with the field, technology, and industry development. Budget pressures were reported to affect the stability of programs, their ability to fill positions, and their plans to expand programs. Some respondents (11%) pointed to problems with obtaining enough qualified faculty.

This question drew 145 responses, some making multiple points (see Table 28).

A sampling of quotes.

We are a newly established program and need to gain an identity and adequate majors. We teach many "service" courses and our majors often don't get a clear sequence of courses nor sufficient work with new technologies.

The internship system is weak in many schools, so many do not have much work experience. And without it, most of them don't take professional preparation courses as seriously as they should. Also, undergraduates lack assertiveness about networking, something they need to do in a tight job market. Finally, writing skills are weak, yet senior faculty complain about teaching basic composition that majors never mastered in upper-division

TABLE 28
Most Significant Issue Facing Undergraduate Programs

Issues Facing Program	n (%) of 145	Frequent Descriptors
Recruitment	35 (24%)	Recruiting and retaining qualified and motivated students
Program weaknesses	28 (19%)	Developing coherent programs, balancing theory and practice, meeting the needs of diverse students, keeping up with the field; bridging academy and industry, providing internships, helping industry link to and understand academic programs
Department relations	19 (13%)	Gaining respect and support of literature faculty, coordinating with other campus programs, connections between major and minor or concentrations in TC
Resources	18 (12%)	Insufficient resources for faculty lines, technology, classes, expansion
Staffing	16 (11%)	Hard to hire qualified faculty, inexperienced teachers in classroom, too few faculty
Economy	13 (9%)	Slow economy, poor job market
Not sure or not applicable	24 (17%)	

courses. Budget issues have impacted class size and access to technology. With larger classes, faculty are overloaded, and adjuncts are not always as skilled as we would like.

We have no funds for recruitment and thus no plan.

The most significant issues facing graduate programs.   When asked to identify significant issues facing the graduate programs in which they taught, 135 respondents cited problems of curricular coherence, balance between theory and practice, or maintaining appropriate emphases in the face of student or institutional pressures. Responses indicated that programs are having difficulty getting faculty lines approved and, when they do, finding qualified candidates. Some programs are facing enrollment declines, and some have trouble attracting students of high caliber. These weaknesses are sometimes compounded by programs being located within departments that are not supportive of the goals of technical communication.

This question drew 135 responses, some making multiple points. Table 29 summarizes the issues mentioned.

A sampling of quotes.

The most immediate issue in our department is the desire by other parts of the department to constrain the growth of the professional writing program and especially new hires and new online courses.

TABLE 29
Most Significant Issues Facing Graduate Programs

Issues Facing Program	n (%) of 135	Frequent Descriptors
Program weaknesses	49 (36%)	Developing coherent programs, balancing theory and practice, meeting the needs of diverse students, keeping up with the field and technologies; bridging academy and industry
Staffing	17 (13%)	Hard to hire qualified faculty, too few faculty
Resources	16 (12%)	Insufficient resources for faculty lines, technology, classes, expansion
Recruitment	15 (11%)	Recruiting and retaining qualified and motivated students
Department relations	13 (10%)	Gaining respect and support of literature faculty, coordinating with other campus programs, connections between major and minor or concentrations in TC
Economy	7 (5%)	Slow economy, poor job market
Not sure or not applicable	39 (29%)	

Trying to become a "real program," which would require a curriculum and the faculty to develop and staff it, neither of which is likely to be provided. We have tried to hire without success.

Our most serious, continuing problem is a workload that is far too heavy for the level of research that we must achieve and program involvement that we must give. A second major problem is lack of funds for adequate program coordination and recruitment and for research and travel. A third major problem is recruitment and retention; our program has shrunk more than 50% over the last three years. We have no funds for recruitment and, thus, no plan.

The most important issue is finding a space in a somewhat nebulous field. Tech comm has grown so much in so many different ways, it seems to lack clear boundaries or definitions—which can make it difficult for students to build a comprehensive sense of the field's possibilities.

I don't currently teach in a graduate program, but here's one problem I see in such programs: Less than 10% of the jobs on the market exist at doctoral-degree-granting institutions (see the 2000 MLA Committee on Professional Employment, Final Report). The vast majority of jobs exist at small schools or regional comprehensive institutions. I don't believe graduate programs are doing enough to help enculturate PhD students to the values, dynamics, and real work of the institutions where they'll be hired. There's an elitism existing in too many graduate programs that refuses to see what is in plain sight. This purposeful blindness hurts the students of these programs by not preparing them for the places/contexts within which they will be working and getting tenure.

Staying relevant and competitive is the main issue. Our program prides itself for being on the leading edge in terms of our courses, our faculty, and our students. We need to maintain that edge. And we need to remain competitive in terms of cost, something we have little control over since we're part of a private institution and don't have much influence over the tuition rates.

Desired directions for the field's future.   The final question of the survey asked respondents, "How do you want to see the field of technical communication evolve, and what implications does your desire for the field's future hold for teaching, research, and theory?" The 123 respondents who answered frequently called attention to the need for the field of technical communication to attain a mature professional identity and, with it, stature as an academic discipline and a field of relevance and value to industry. Some suggested broadening the activities of the field to embrace new areas and technologies of communication. While some argued that the field must remain practical and applied, especially with regard to producing research with direct industry applications, others insisted that it is important to grow theoretically as an academic discipline (see Table 30).

TABLE 30
How Respondents Want to See the Field Evolve

How Should Technical Communication Evolve?	n (%) of 123	Frequent Descriptors
Gain identity and professional stature	32 (26%)	Name recognition, program definition, self respect, professionalism
Broaden the kinds of writing and communication work we do	24 (20%)	New media, design, technology, medical and scientific communication, communication more broadly, civics, service and client learning
More connection to industry	21 (17%)	Stay pragmatic and pursue applied research in industry settings; close the gap of academia and industry
Deeper theoretical and research grounding	20 (16%)	More and more useful research; deeper connections to theory, stronger humanities base, balance between theory and practice, more connections to critical and cultural theory
Pursue interdisciplinary connections	13 (11%)	Through research, double majors, campus connections to management and computer science, journalism
Better positioning within the university	11 (9%)	Closer connections to writing and literature programs, wac, increase profile and administrative support
Keep up with changes	6 (5%)	Technology, interdisciplinary research
No answer	7 (6%)	Too complex

A sampling of quotes.

I think that tech comm needs to get much smarter about its position (and possible positions) in the world. We train our students like head-down, task-focused engineers when what they need—and can have—is a broader business sense that situates their work and gives their interdisciplinary strengths more leverage. To me, knowledge management is a good start, but we need public policy, profit and nonprofit management theory, and rhetoric of technology if they're ever to get in a position to really push the humanistic ethics we're trying to give them.

I think we have social responsibilities as technical writing teachers/professionals to make our students clear and creative thinkers. We should not be slaves to the technology.

I will begin my PhD in Professional Writing this fall, and I am passionate about the field. My main concern at this point is that there is too much emphasis on "publishing" in academic journals and not enough on interacting and publishing with the very businesses that we are discussing every day in the classroom.

We need to research the actual value of good communications and quantify that value in a way upper management can understand. Then, we need to instill a degree of professionalism in our students that will cause them to act in a manner that suggests they are a critical part of their companies' success.

I like the movement I see in the direction of ethics, civic rhetoric, and public policy. We need to be good citizens as well as skilled technical communicators.

We should be concerned and cautious about our potential role in consumerizing the entire world. We can have important, responsible roles to play in making the world a better place for people, not simply for corporations, for the environment, not simply for the exploitation of resources, but at this point we are not thinking in those terms. Teaching, research, and theory need to include such areas as postcolonial theory, social theory, and globalization studies.

I'd like to see the field develop a clearer sense of its identity, history, and significance. Everyone understands without question what an engineer is—but my students have often commented on how difficult it is to explain to friends and parents what tech comm is and what they do. The problem lies in our inability to define ourselves, which I think is tied to two deeply held beliefs in the field: that we should be the anonymous servants of technology and its users, and that because the communications we create are ephemeral, we have no real or at least no recognized history. These are values no engineer (or doctor, lawyer, teacher) would ever subscribe to. I'd like to see the field develop more confidence in its significance at the center of knowledge creation, as well as promote an awareness of this significance in academic and corporate communities. To do so, we should focus on teaching students (in both service courses and disciplinary ones) not just how to fulfill generic conventions, but how to manage information and people, how to make wise and ethical decisions about communication, and how to bend technology to the task of communication. We should also help students recognize the long and complex history of communication about technology, science, and business; awareness of the field's development is central to understanding its dynamics and its relationship to cultures, whether local, national, or global.

A BRIEF DIALOGUE OF CONCLUDING OBSERVATIONS

To wrap up this data-intensive text, we decided that each of us would write a statement highlighting what we found most interesting about the survey results. We hope our concluding comments will encourage readers to end this article by extending it—by adding their own commentary on the survey to those posted at the ATTW Web site in the *TCQ* Comments forum.

Dayton: Unexpected Unity within Diversity

The survey results show, with striking statistical clarity, how we divide pretty evenly into three groups defined by distinct disciplinary emphases:

- English studies and literature
- Technical/professional communication, with and without rhetoric
- Composition studies and/or rhetoric

When you look at the titles on our highest academic degrees, roughly even numbers of us divide into the first two groups and about one-quarter fall into the third group. The first group, to my mind, represents the earliest layer of our archeology as a field. The second group identifies the disciplinary rubric that has become dominant in the degree-granting programs that are specifically ours, although they are often dependent on faculty whose degrees put them in the third group.

Do these groupings mean anything? I think they do. I think they are the root of our perpetual identity anxiety ("crisis" seems too strong a word for a condition so chronic). Those who identify primarily with composition studies and/or rhetoric are likely to view the field more broadly than those who identify primarily with technical/professional communication (with or without rhetoric as an addendum). And even within the second group, the need to include both *technical* and *professional* as modifiers of *communication* points to a more subtle but still consequential difference in perspective on the boundaries of our knowledge domain—or, at least, how best to present those boundaries to the world at large.

Awkward as it is, the phrase "technical and professional communication" is the name that binds the greatest number of us together, as shown by the data on preferences for what to call our field (Table 22). I think we would be smart to hang onto that disciplinary rubric, adding "rhetoric" for our doctoral programs or modifying the phrase slightly to appease the gauntlet of committees we have to run when proposing new doctoral degrees, as East Carolina University is currently doing with its proposed PhD in Technical and Professional Discourse.

What most surprised me in the survey results was the strong and growing proportion of ATTW members who are working in programs either dedicated to technical and professional communication (TPC) or offering a TPC concentration within an English or writing major. When I entered the job market in the fall of 2000—anticipating the completion of my PhD in 2001—I discovered that most job announcements mentioning technical and business writing on the MLA job list were in English departments with no major or concentration in TPC.

What proportion of college or university teachers of TPC courses are ATTW members? A study looking into that question would seem worthwhile, for if we are not a majority of that larger population, we should be. How might we reach out more effectively to TPC teachers who have not yet seen the advantages of ATTW membership or to those who joined for a time and then let their membership lapse? Surely, we have a lot to offer the many lone TPC teachers in colleges and universities. Are we providing them with the support that could help them get more satisfaction out of their career and, perhaps, motivate them to grow new programs in TPC?

Bernhardt: What Has Changed and What Hasn't

The survey demonstrates professionalization of the field—with many respondents working from within established programs in technical/professional communication at both undergraduate and graduate levels. To a great extent, we seem successful at reaching this audience and probably less successful in serving other audiences, such as the large numbers who teach in small or community colleges or who are engaged primarily in service teaching. Membership and leadership profiles suggest a balanced field, with even numbers of newer members working alongside those who have been in the field for years.

The survey also underscores the limited reach of ATTW as primarily an organization serving university teachers in the United States. There is still a lot of program development going on, with high numbers indicating they are working to define their programs, gain respect of their departments and schools, and fully professionalize their work. Several respondents suggested that ATTW should do more to involve a wider range of people in its leadership and in its issues. This is a useful warning: to not become too insular.

I'd be disinclined to suggest a realignment of the conference, since so many members are closely allied with NCTE/CCCC and attend the CCCC annual meeting. It is interesting that although most closely aligned with CCCC, members read *Technical Communication*, the STC journal, in higher numbers. We'd be foolish to engage in a name debate, since there are so many opinions and so little consensus that it would prove futile. One surprise in the data was the number of alternative formulations for who we are, coupled with relatively few people agreeing on a governing term. I tend to agree with David that "Technical and Professional Communication" best captures what we do and communicates it well to those outside the field.

There seems to be high appreciation and use of the journal and its yearly bibliography. While I have considered the listserv to be the primary means of member communication, it is striking that relatively large percentages do not read the list or participate only occasionally. They do, on the other hand, go to the Web site to get information, especially about the conference.

The open-ended questions give us a really useful portrait of what we believe we should be teaching and what we should be working on doing better. Programs would be well served to evaluate themselves (as programs) and their students against the list of important skills as outlined here. Table 30 offers a substantial "to-do" list for the field. There's plenty to work on.

I had been hoping that the field would not be affected too strongly by the economic downturn, that programs and their graduates would continue to experience good fortune. I think the situation is darker than I'd hoped. Compounding the dark sense of the economy are deep-seated and unresolved conflicts over identity in the field—who our students are, who we want them to be, what they want to be. This seems a central tension that many faculty are working through.

David Dayton is an assistant professor in the Department of Humanities and Technical Communication at Southern Polytechnic State University in suburban Atlanta. He has published articles in *JBTC*, *JTWC*, and *TC*, and he authored the chapter on electronic editing in the third edition of Carolyn Rude's *Technical Editing* (Allyn & Bacon/Longman, 2002). His doctoral dissertation in Technical Communication and Rhetoric at Texas Tech University won the CCCC Outstanding Dissertation Award in Technical and Scientific Communication in 2002.

Stephen A. Bernhardt is the Kirkpatrick Chair in Writing and Professor of English at the University of Delaware. He is past president of ATTW and of CPTSC.

STC's First Academic Salary Survey, 2003

Sandi Harner
Cedarville University

This article reports United States salary data from the April 2003 survey of Society for Technical Communication members who identify themselves as educators. It provides analysis of salary data based on type of institution, rank, tenure status, experience, education level, sex, and age. It also reports on benefits, administrative responsibilities, job satisfaction, and program size.

In April 2003, the Society for Technical Communication (STC) launched its first salary survey for full-time faculty members teaching in technical communication programs at two- and four-year colleges and universities in the United States. Members from universities with graduate programs are incorporated into the four-year group. Adjunct faculty members were not surveyed. This survey looks at compensation as well as at other information specific to this group of educators.

STC sent this survey to a total of 2,163 members, all of the STC members who identified themselves as educators. Industry trainers may identify themselves as educators, but their responses were omitted from the results reported here. Of the members who received the survey, 269 responded (12.4%). Of those responses, 163 (7.5% of the total) contained sufficient information to be included in the survey. This response rate is better than the percentage sounds because so many industry trainers were included in the original mailing. By comparison to this sample size of 163, the 2003 survey of ATTW members generated a sample size of 228. (See the article by Dayton and Bernhardt in this issue.) Members received their salary survey forms by mail and returned them on paper copy.

Results of this survey are presented in Table 1, modeled after the STC annual salary survey of all members (see www.stc.org).

In addition to salary, full-time academics in technical communication receive the benefits shown in Table 2.

In addition to seeking information on salary and benefits, the STC survey queried respondents about administrative responsibilities, job satisfaction, and program

TABLE 1
2003 Academic Salary Survey: Annual Wages—U.S. (U.S. dollars per year)

Grouping	Base	Mean	10%	25%	50%	75%	90%
All Respondents	163	$55,015	$36,000	$43,300	$51,000	63,000	80,000
Type of Institution							
4–year	143	55,020	36,000	43,000	50,000	64,150	80,000
2–year	20	55,000	—	45,000	52,000	60,000	—
Academic Rank							
Professor	47	71,360	—	58,500	69,800	80,000	—
Associate professor	35	58,840	—	51,000	55,320	64,450	—
Assistant professor	51	44,220	37,900	40,000	45,400	47,280	50,000
Instructor	26	42,610	—	30,000	40,000	52,000	—
Tenure							
Tenured	76	64,780	49,000	53,020	60,000	70,000	90,000
Nontenured	83	45,620	32,000	38,000	44,100	48,400	61,700
Years of Teaching							
Fewer than 6 years	33	46,170	—	40,000	45,000	50,000	—
6–10 years	33	45,500	—	40,000	45,000	50,000	—
11–20 years	49	54,700	—	45,000	51,900	52,500	—
21–40 years	48	68,000	—	55,700	65,000	78,000	—
41 years or more	—	—	—	—	—	—	—
Years in Industry							
Fewer than 6 years	96	55,950	36,500	44,000	52,500	66,000	80,000
6–10 years	32	54,320	—	43,500	50,000	57,000	—
11–20 years	27	52,140	—	41,500	51,090	60,360	—
21–40 years	8	56,430	—	—	—	—	—
41 years or more	—	—	—	—	—	—	—
Education Level							
Master's degree	40	47,530		36,000	46,000	60,000	—
Doctorate	122	57,520	40,000	45,000	54,000	67,000	80,000
Sex							
Male	73	61,120	43,000	47,280	56,400	71,000	90,000
Female	90	50,070	34,900	40,000	47,000	60,000	68,000
Age							
20–29	5	35,000	—	—	—	—	—
30–39	36	46,760	—	43,250	47,000	50,000	—
40–49	29	48,360	—	39,000	47,000	55,000	—
50 years and over	92	61,600	40,000	47,300	59,950	71,000	90,000

Note. Copyright 2003 by the Society for Technical Communication (STC). Reprinted with permission of STC.

Base = The total responses in a given category. Mean = The value computed by averaging the tabulated responses. 10% = Ten percent of the responses were below this value; 90% were above this value. 25% = Twenty-five percent of the responses were below this value; 75% were above this value. 50% = Fifty percent of the responses were below this value; 50% were above this value. (This point is also called the *median*.) 75% = Seventy-five percent of the responses were below this value; 25% were above this value. 90% = Ninety percent of the responses were below this value; 10% were above this value.

TABLE 2
Employee Benefits

Dental insurance	75%	Professional society dues	21%
Disability insurance	7%	Retirement/pension	90%
Health insurance	94%	Seminar/conference reimbursements	76%
Internet access	78%	Tuition reimbursement	43%
Life insurance	72%	Other	3%

size. Perhaps administrators are disproportionately represented in this sample because 76% have administrative responsibilities, but 94% reported no extra compensation for their administrative roles. This sample probably represents a disproportionate number of advanced professors among academics as a whole because more than half are over age fifty, and 60% have been teaching for eleven years or more.

Respondents serve in the following roles:

Dean	4%
Department chair	16%
Director of TC	56%
Advisor of STC student chapter	24%

At least on the basis of this relatively small sample, academics in technical communication are generally satisfied with their jobs, with only 8% reporting dissatisfaction. They reported these results in response to the prompt "What is your level of satisfaction with your present position?"

Very satisfied	40%
Satisfied	52%
Dissatisfied	6%
Very dissatisfied	2%

These figures are confirmed by their reasons for choosing this profession. Respondents could choose more than one reason, but 55% chose "Enjoy type of work" as their primary reason.

Earn good income	15%
Earn respect of others	23%
Enjoy type of work	91%
Enjoy working conditions	70%
Fell into it	22%
Make contributions	60%
Use my talents	54%
Other	15%

Technical communication programs remain small. The size of the program may impact salary negotiating power. Survey respondents reported the number of full-time faculty teaching in their technical communication program:

1	11%
2–3	30%
4–5	28%
More than 5	31%

STC has proved valuable through this first effort at a survey of academic salaries. We learned many things as we created the survey instrument and looked at the resulting data. We were not able to draw conclusions in some areas that are of interest to all academics. Perhaps next year's survey—or other new research endeavors—could begin to collect important data on the following issues:

- Faculty teaching loads
- Salary compression
- Male and female salaries of those who are not deans or department chairs
- Comparison of salaries based on program academic location (English, technical communication, engineering), geographic location (urban, rural), university type (community college, liberal arts college, state university, private university), and program type (dedicated degree programs, service courses)

Overall, we have discovered some important information that should benefit all academics. We need to remember, however, that this is not scientific research. This data reflects information given on the salary survey of STC members who identify themselves as academics. We have no way to verify the information but we rely on the honesty of those who return their surveys. The results are also based on a limited sample of academics and should be validated in a follow-up survey. I thank STC for sponsoring this survey.

Sandi Harner is a professor at Cedarville University. She designed and is currently Director of the Technical and Professional Communication Program. She co-authored a book entitled *Technical Marketing Communication*, which was published by Allyn & Bacon/Longman in 2002. She currently serves on the STC Board of Directors in her fifth year as Assistant to the President for Academic and Research Programs.

The Academic Job Market in Technical Communication, 2002–2003

Carolyn Rude
Virginia Tech

Kelli Cargile Cook
Utah State University

Analysis of the academic job market in 2002–2003 reveals that 118 nationally advertised academic jobs named technical or professional communication as a primary or secondary specialization. Of the 56 in the "primary" category that we were able to contact, we identified 42 jobs filled, 10 unfilled, and 4 pending. However, only 29% of the jobs for which technical or professional communication was the primary specialization were filled by people with degrees in the field, and an even lower percent (25%) of all jobs, whether advertised for a primary or secondary specialization, were filled by people with degrees in the field. Search chairs report a higher priority on teaching and research potential than on a particular research specialization, and 62% of all filled positions involve teaching in related areas (composition, literature, or other writing courses).

Over the past decade, development of new academic programs in technical and professional communication and expansion of existing programs have increased the demand for faculty with credentials in the field. However, the number of graduates emerging from degree programs in the field has been inadequate to meet the demand. Failed searches are common, opportunities for lateral or upward academic movement create retention problems, and a limited number of applicants for any position complicates selection.

An imbalance between demand for faculty and supply creates multiple problems and can compromise the development of the field overall. Programs cannot develop as planned, positions may be filled by people with little preparation and interest in the field, and research may suffer if faculty positions are held by people unfamiliar with the methods of research and research questions in this field. Doctoral programs would like to meet the need, but they may not have the capacity on their own faculties to increase enrollment. New doctoral programs may develop to help, but they, too, may not have sufficient faculty for robust programs. Furthermore, the pool of applicants eager for doctoral study may be limited. Comments by program directors at the 1999 meeting in Santa Fe of the Council for Programs in Technical and Scientific Communication (CPTSC; see Brumberger; Gurak; Lay; and Rude) note the difficulties of recruiting students for doctoral study.

Predicting future needs on the basis of current demand is an uncertain and precarious task. Projecting the future on the basis of the present presumes continued growth and development both academically and in the workplaces that hire the students whom the faculty teach. A faculty member may have a career of thirty years or so, and unless the program grows, a position in it, once filled, may not open again for thirty years. (New faculty may occupy the position in the case of movement of faculty from one university to another, but in such a case there is no net gain in the number of positions overall, just a shifting of the persons who occupy the positions.) It takes three to five years to develop a graduate with a PhD, and planning admissions for a job market that many years prior to the job search can be risky. Growth of academic programs and the parallel demand for new faculty seem tied to growth of the role for technical communicators in the corporation (the dominant employer of graduates from undergraduate and master's programs), and academics have limited impact on that demand.

One way to enhance the ability to predict and plan is to develop data on the academic job market. Right now, such knowledge is based primarily on anecdote. Usually the anecdotes focus on failed or difficult searches, creating the impression of a serious imbalance between demand and supply. Furthermore, anecdotes do not distinguish among the various types of academic positions. Unlike other subjects taught in all colleges (mathematics, biology, composition), technical communication may not be offered at all. When courses are offered, they may be taught as service courses only or within a dedicated program at the undergraduate or graduate level. Teaching requirements for positions that identify the specialization may include the occasional technical communication course in an assignment that emphasizes literature or creative writing. It would also be useful for prospective and current doctoral students to understand the varieties of jobs available and the requirements for them. It would be useful for search committees to understand how their own searches compare with others. All would benefit from knowing what specializations within the field as a whole are in demand and how the demand for specialists compares with the demand for generalists who can teach other courses (usually literature, creative writing, or journalism) along with technical or professional communication.

We conducted this study in an effort to describe the job market systematically and to provide a baseline for further study. We have identified the 2002–2003 academic positions for which technical or professional writing was either a primary or secondary requirement, have classified these positions according to type of academic program and specialization, and have contacted search chairs to learn the status of the search as of late April 2003 and the qualifications of the person hired. We report on search strategies and selection criteria. We have also tried to learn about teaching responsibilities and programs where these positions are advertised. We report our findings here, cautioning that they describe just one year in the academic job market that may or may not be a typical year and that may or may not predict the future.

METHODS

To identify the positions for which preparation in technical and professional communication was a primary or secondary requirement, we read all the ads in the Modern Language Association (MLA) Job Information Lists for October and December 2002 and February 2003. In addition, we reviewed the positions listed at the Web sites of the ATTW (www.attw.org) and the CPTSC (www.cptsc.org). We listed all the ads that named technical, professional, business, or scientific writing. In addition, we listed ads that less directly implied an interest in technical or professional communication, including ads for specializations in writing across the curriculum, Web site design, or visual communication. We identified a total of 133 possible positions, 15 of which we later eliminated after talking with departments and learning that positions involved no teaching in the field, leaving a total of 118. We counted by position, not by institution. If one college or university advertised multiple positions, even with the same description, we counted each position separately. We sorted these positions according to whether credentials in technical and professional communication represented a primary or a secondary specialization, identifying 60 primary and 58 secondary positions.

We contacted the search committee chair (or in some cases a human resources staff member) for each position by phone or e-mail to ask questions about the status of the search, qualifications of the person hired, salary, program description, teaching responsibilities, and selection criteria. If the search chair indicated that the position would not require teaching in technical or professional communication, we eliminated the position from our initial list. From the final list of 118 positions, we succeeded in interviewing the search chairs of 95 positions, for a response rate of 81%. Our rate for the positions where technical communication was the primary requirement was higher: 56 of 60, for a response rate of 93%. We do not report on specific searches but rather report data in the aggregate. Our methods and interview questions were approved by the Institutional Research Boards at both of our universities.

Although our response rate suggests a high degree of reliability in our results for the positions advertised through MLA or CPTSC, our method was inadequate for identifying a complete list of positions at community colleges, which often do not advertise through MLA, ATTW, or CPTSC, and may be likely to hire from a pool of local applicants. Likewise, most of the positions advertised through our sources were for tenure-line, professorial positions, not for lecturer or instructor positions. In these regards, our description of the academic job market is incomplete. It focuses on professorial, tenure-line positions.

In addition, based on our reading of the ads, several schools appeared to advertise primary positions for technical or professional communication specialists, but as we learned upon interviewing the hiring departments, the individuals hired would teach a combination of classes, sometimes with only one technical or professional writing course assignment per year. Although we did not change the position's classification from primary to secondary in our data analysis, we acknowledge that this decision may skew the description of the job market somewhat. Nevertheless, we felt it was important to read the position advertisements separately from the information we gathered in the interviews because it gave us a picture of what departments said they wanted (their advertisements) and what they actually got (their hires). Maintaining our classification based on job ads may also alert graduate students about varied expectations in positions described in similar ways. The confusion about what a specialist in the field may be expected to do also speaks to difficulty of defining individuals who hold degrees in technical and professional communication and relates to the naming issue and identification of the field (i.e., What are specialists in technical or professional communication, and what do they teach and research?).

We concentrate in some of our analysis on the 60 positions for which technical or professional writing is the primary specialization. The 58 positions for which technical or professional writing is a secondary specialization usually list composition and rhetoric as the primary specialization, but 2 list literature, and 22 are seeking generalists in English who can teach a variety of literature and writing courses.

First, we describe the jobs that were advertised, including qualifications requested, rank, and types of program (specialization for majors or service courses) and levels taught (graduate or undergraduate). Then we describe and analyze the search and selection process at the institutions that succeeded in hiring. We also describe the teaching responsibilities of the candidates sought and salaries offered. Finally, we offer some comments based on the data.

THE JOBS: SPECIALIZATIONS, RANKS, PROGRAMS, COURSES

In this section, we analyze the advertised positions: the specialties they sought and the ranks they requested. We also report how departments describe themselves and what courses they expect new hires to teach.

Specializations and Ranks

Approximately half of the positions (51%) that we identified through the job ads named technical or professional communication as a primary qualification. Not all of the positions for specialists, however, involved full-time teaching in technical communication, with one extreme requiring the person hired to teach three literature courses and one technical communication course each semester. But it is perhaps an affirming sign that a department would seek a person with preparation in technical communication to teach literature as well rather than the alternative of hiring a litera-ture specialist assuming that that person could teach technical communication.

The other half of the positions we identified that welcomed or required prepara-tion in technical communication identified this field as a secondary specialization (49%). For these kinds of jobs, degrees in rhetoric or English with some courses in technical communication would provide good qualifications. Not surprisingly, given the preference for generalists who can wear multiple hats, most of these po-sitions are in undergraduate programs that offer service courses, undergraduate minors, or undergraduate concentrations, but 3 of the positions are at universities with doctoral programs in some aspect of writing.

The positions advertised for a variety of ranks from teaching fellows and lectur-ers to full professors. By far, the most common rank advertised for both primary and secondary positions was assistant professor (67%). The least commonly ad-vertised position was full professor (3%). Table 1 summarizes additional findings about all positions advertised, focusing on specializations and ranks sought.

Program Varieties and Courses

Programs advertising for a primary or secondary specialist in technical or profes-sional communication vary widely in makeup, although they are overwhelmingly housed in English or humanities departments (given a variety of names, including General Studies, Integrated Studies, Arts and Humanities). One position each was advertised for Business Administration or Engineering departments, and two Com-munications departments sought candidates in the field. Only two hiring depart-ments were dedicated to the field according to their names: Professional and Techni-cal Communication and Institute for Technical and Scientific Communication.

TABLE 1
Jobs Advertised by Rank and Primary or Secondary Specialization
in Technical Communication, 2002–2003

	Rank					
Specialization	Assistant	Assistant/ Associate	Associate/ Full	Lecturer/ Instructor/ Fellow	Rank Open or Undeclared	Total
Primary	45	3	3	7	2	60
Secondary	34	6	1	14	3	58
Total	79 (67%)	9 (8%)	4 (3%)	21 (18%)	5 (4%)	118 (100%)

Within these departments were eighteen different program varieties, which we were able to place into these eight general categories, depending on the courses and degrees offered:

1. Service courses only
2. Service courses and undergraduate majors and/or minors
3. Service courses, undergraduate majors and/or minors, and master's degrees
4. Service courses, undergraduate majors and/or minors, master's degrees, and doctoral degrees
5. Service courses and either a master's degree or certificate and/or a doctoral degree
6. Undergraduate majors and/or minors only
7. Undergraduate majors and/or minors and master's degrees
8. Undergraduate majors and/or minors, master's degrees, and doctoral degrees

Within these varied programs, teaching assignments varied as well, ranging from technical or professional writing only to combinations including rhetoric, composition, literature, communication, technical journalism, creative writing, and creative nonfiction at both the graduate and undergraduate levels. Some ads for specialists targeted research or teaching expertise in multimedia, editing, visual rhetoric, science and medical writing, research methods, or knowledge management. (Areas of specialization are discussed in more detail in the article's next section.)

Overall, the 118 position announcements in 2002–2003 reveal a varied and open market for job seekers. Although this market was primarily focused on entry-level academic positions (from lecturers to assistant professors), it also offered several opportunities for movement to higher ranks. Job candidates could choose from programs that offered service courses only to those offering technical or professional communication courses to all levels of students (service, undergraduate, and graduate).

SEARCH METHODS

In departments of English, the academic home of most programs in technical and professional communication, the traditional hiring pattern is to advertise in the fall job list of the Modern Language Association (MLA) and to interview candidates at the December MLA meeting. After the interviews, departments may select one or more candidates for a campus visit. The job offer follows. When it is difficult to hire qualified faculty members, departments may try to gain a competitive edge by modifying these search patterns and by making strategic choices at each stage. In this section, we narrow our focus from the 95 positions we surveyed to a smaller group: 36 primary professorial (tenure-line) positions that were successfully filled this year (representing thirty-three colleges and universities, three of which filled 2 positions). This total eliminates 6 lectureships from the total of 42 filled jobs for which technical communication is a primary qualification, and allows us to concentrate on search methods and strategies for filling professorial tenure-track positions, by far the most commonly advertised and filled.

The Job Ad: Targeted or General?

The job ad requires strategic thinking. Should it target particular specializations within technical and professional communication or define candidate strengths broadly? A specialization, such as environmental writing or multimedia design, eliminates some candidates, but it may attract the candidates who would be the best match for the department. One department chair happily reports to us success in a very targeted search for a narrowly defined position. This chair acknowledges that the ad diminished the number of applicants but not the quality of those who were suited for the job. Of the 36 professorial (tenure-line) positions advertised and filled in 2002–2003 for which technical or professional communication was the primary specialization, job ads for 17—or 47%—advertised for generalists, and 19—or 53%—for specialists, but some of the specializations were broadly defined ("discourse and culture"), and at least three job ads included seven to ten specializations, putting them more in the broad rather than targeted category. The sample size is too small to support firm conclusions on the matter of targeting or generalizing, but of 14 jobs unfilled in the "primary specialization" category, 12 identified a specialization that may have discouraged some applicants (or there could have been other reasons for the failed searches). For example, if an ad names multimedia writing, a candidate whose strengths are elsewhere in the field may not apply.

The numbers support two conclusions: (1) The field and programs are generally not big enough to support the hiring of people with specific areas of interest, at least at the hiring stage; and (2) because of the small pool of candidates, many departments search broadly. These departments seek the best candidate available and hope the candidate is able to teach the necessary courses while pursuing special interests in research. The good news of undefined specializations is that faculty may be able to cross boundaries rather than specializing too narrowly, and over the course of a career, the breadth may invigorate both teaching and research. The bad news is that the field remains amorphously defined to department chairs and deans, with the particular areas of expertise unrecognized. Faculty may seem to be interchangeable. They may have less academic identity within an English department than, say, a colleague who specializes in seventeenth-century British poetry. This lack of definition may also hinder programs' abilities to define hiring needs and argue for new or replacement tenure-track lines: Without identified specializations, how does a program demonstrate that it lacks in a specialized area of expertise? This issue of specialization is more significant for doctoral programs than for undergraduate programs, because faculty must be prepared to direct dissertations in particular areas, but it is also significant for the field as a whole. Any field needs to develop its body of knowledge through research, and if faculty positions are filled by people with general preparation in English studies rather than specialized preparation in technical communication and rhetoric, the research may be limited. And as the analysis of salaries reveals, a generalist position is likely to carry a lower salary than a specialist position.

Of the nineteen ads that did name a specialization, sixteen named some variation of technology (Web design, multimedia, digital rhetoric). The next most frequent specialization was science or scientific writing (five ads). Visual communication, research methods (including usability testing), editing, and health or medical writing each got three mentions. These specializations seem to be teaching fields rather than desired research areas. Job candidates would be wise to prepare to teach in these fields no matter what is the specialization of their dissertation research.

Advertising the Job, Identifying Potential Applicants

Although an announcement in the MLA Job Information List will probably be seen by most applicants, advertisements in multiple markets are common. Both the ATTW and CPTSC maintain job lists at their Web sites (www.attw.org; www.cptsc.org). Announcements of positions on the listservs of these organizations (attw-L; cptsc- L) are also common, as are announcements on other listservs. Some departments use direct mail to graduate advisors. Many departments provide information on their jobs at their Web sites.

Informal methods of identifying candidates and inviting applications may be as effective and necessary as these formal announcements. The field is small, and faculty and job candidates can get to know each other fairly easily through the professional associations. Courtship of candidates may begin a year or more before a job is announced as faculty make a point of getting to know graduate students who attend various national or regional meetings. The "Special Interest Group" meeting that ATTW sponsors at the meeting of the Conference on College Composition and Communication (CCCC) each spring and the research network forum at the spring ATTW conference bring together graduate students and faculty in small discussion groups. The fall meeting of CPTSC, usually in October, has become a good place for networking about jobs. CPTSC encourages graduate student participation, and some job candidates are usually among the presenters. Program directors may talk with each other at the conference about jobs and candidates, and potential matchmaking begins. Such conversation may also take place online. Having identified promising candidates, faculty may invite an application by e-mail, phone, or direct mail. These methods obviously favor departments whose members are active in professional associations, often departments with established programs. As such departments gain an edge through these recruitment efforts, new and limited programs may find it increasingly difficult to hire specialists. Of course, these aggressive methods of identifying and recruiting applicants do not explain the whole story of a successful search. Of the 36 jobs filled, 17 were at colleges and universities with regular participants at the conferences of ATTW and CPTSC.

Although the MLA interview remains common in the successful search, a number of schools bypassed MLA in 2002–2003, using telephone interviews and campus visits instead. An MLA interview was part of the search for 20 of the 36 successful searches. A telephone interview substituted for the MLA interview in all but one of the remaining searches. But nearly half of these departments skipped MLA, putting their recruitment dollars into campus visits. For each position, a range of one to five candidates visited campus, with distribution as shown in Table 2.

In addition to these search strategies of marketing the job and identifying candidates, some schools may modify the search schedule. If departments bypass MLA, they are not subject to the MLA requirement that no job offers be made before or during the December meeting. Some departments have offered jobs before MLA, sometimes with the condition that the candidate accept before MLA and withdraw from all MLA interviews. In addition to preempting the candidate's search, this method sometimes offers an advantage by enabling a department to complete the appointment before a budget freeze takes place.

Although these advantages may assist departments to "beat the competition" and reduce the effects of a tight economy, MLA participation does have its benefits for job candidates, especially those who are on the market for the first time. On the one hand, participation in the MLA job market may slow a department's job search, but on the other, MLA offers protection to job candidates by requiring departments to give candidates sufficient time before accepting or rejecting a position, and job candidates may prefer the freedom to explore options unless they have themselves targeted a particular college or university. The situation is most difficult when job candidates apply for some positions outside the MLA market and others within it. Furthermore, substituting campus visits for MLA interviews has time and financial implications as well. While campus visits allow search committee members to have a holiday break, campus visits may be more expensive, depending on the number of candidates brought to campus. They can also be more

TABLE 2
Campus Visits in Departments

Number of Departments	Number of Campus Visits per Search
8	1
8	2
13	3
6	4
1	5

time consuming for search committee members who spend two to three days with candidates (as opposed to thirty minutes to an hour in MLA interviews) and for candidates (especially for those who are working to finish a dissertation by May).

SELECTION CRITERIA

We asked search committee chairs about the candidate qualifications that influenced job offers. The numbers that follow are based on the 36 professorial positions that were filled. Chairs of these search committees could say realistically what standards were in fact applied in selection. Table 3 is arranged from highest to lowest in value as marked by search chairs naming the qualification "very important" or "important." (Occasionally a respondent did not comment, and the totals regarding recommendations and work experience in technical communication do not add up to 36.)

These results confirm the analysis of the job ads: The particular specialization is less important to employers than demonstration of teaching and research potential. Employers may be happy to hire a qualified person whatever the dissertation area, and hope that the person will meet the teaching needs. Or they may be more interested in coverage of courses than in the subject of the research that their new professor will do. Or they may perceive that technical communication is itself a specialization whose inquiries are undifferentiated. (This possibility is supported by the fact that many positions are filled by people with degrees in English studies rather than in technical communication and rhetoric.) If the field grows and if the

TABLE 3
Qualifications at Professional Ranks Valued by Hiring Institutions ($n = 36$)

Qualification	Very Important	Important	Not Important	Total "Very" or "Important"
Teaching experience	30	6	0	36
Job talk, teaching demonstration	26	10	0	36
Prior publication	16	19	1	35
Prior conference presentations	13	21	2	33
Dissertation completion date	24	8	4	32
Recommendations	15	17	1	32
Reputation of the degree granting institution	8	23	5	31
Dissertation topic	11	19	6	30
Research method	6	21	9	27
Reputation of recommender	4	22	10	26
Work experience in technical communication	8	15	5	23

ratio of candidates to jobs increases, search committees might refine their criteria and rate the specialization higher.

THE HIRES: QUALIFICATIONS, RANKS, TEACHING, SALARIES

In this section, we describe the individuals who were hired into the 69 positions from our survey that were filled. Of these 69 positions, 42 (61%) were primary positions, and 27 (39%) were secondary positions. Specifically, we report the hires' previous employment backgrounds, the ranks at which they were hired, the degrees hires hold, and the teaching versatility of these different degrees. We also discuss our findings about their negotiated teaching loads, course assignments, and salaries.

Background and Ranks

In the positions primarily focused in professional and technical writing, hires were previously studying or employed in three different areas: graduate schools, academe, or industry. As might be expected, over half of the primary positions (24, or 57%) were recent graduates; of these 24 graduates, 3 (7%) marketed themselves as having extensive industry experience. One-third (14, or 33%) of this year's hires were full-time academics moving to other positions; only 1 hire (2%) came directly from industry. The remaining 3 hires (8%) were not identified by previous employment. Similarly, graduates comprised the majority of secondary hires (17, or 63%) with academics filling 8 (30%) of the other hires. Only 1 individual (3%) employed in industry was hired into a secondary position, and 1 (3%) was unidentified by previous study or employment.

Departments typically filled their positions at the ranks advertised. Table 4 reports the departments' advertised ranks and the ranks at which they hired. It also shows that nearly three- quarters (74%) of the hires in the 2002 market were assistant professors, whether departments were searching for someone with a primary or secondary specialty in technical or professional communication. Teaching fellows, instructors, and lecturers filled 13 (19%) of the advertised positions, but associate and full professors filled only 2 (3%) of the openings. Table 4 also shows the movement at the assistant professor rank. People already holding academic positions (that is, past graduate school and employed full-time) accepted 12 of the 29 assistant professor positions in the primary specialization category. This number is a marker of retention problems. Although we do not know exactly what their previous positions were (some may have been temporary positions or instructorships), the movement of 12 people surely disrupts the programs where they were working and may point to vacancies in the future. It represents at least in part a shifting of the location of vacancies, not really a filling of positions considered from the perspective of the field as a whole.

TABLE 4
Advertised vs. Hired Ranks for Primary and Secondary Positions and Previous Employment Background

Advertised Rank	Hired Rank	Graduates		Academics[a]		Industry		Unidentified		Total
		PRI	SEC	PRI	SEC	PRI	SEC	PRI	SEC	
Teaching fellow	Teaching fellow	5	0	0	0	0	0	0	0	13
	Instructor/lecturer	1	5	1	0	0	0	0	1	
Open rank	Assistant professor	0	0	1	1	0	0	0	0	51
	Assistant professor	17	11	10	5	1	1	0	0	
Assistant or associate professor	Assistant professor	0	1	1	1	0	0	1	0	
Assistant professor	Unreported rank	1	0	0	0	0	0	2	0	3
Assistant professor	Associate professor	0	0	0	1	0	0	0	0	2
Associate or full professor	Full professor	0	0	1	0	0	0	0	0	
Totals		24	17	14	8	1	1	3	1	69

[a]22 of the 69 people hired in 2002–2003 moved from the other academic positions.

Degrees Held

Hires who filled primary positions held six different degrees or degree specialties. One position (3%) was filled by an individual with an MA/MS degree, and 3 positions (7%) were filled by MFAs. Individuals with PhDs filled the other 38 positions, although their doctoral focus or emphasis varied. Five (12%) had earned PhDs in varying fields both inside and outside of English departments, such as American Studies, Cultural Studies, Instructional Technology, and Linguistics; 6 (14%) were unidentified by their specialties; 7 (17%) held PhDs in English literature with some coursework in professional or technical communication; 8 held degrees in Rhetoric and Composition; and 12 (29%) held PhDs in professional and technical communication. While PhDs in technical or professional communication were the most commonly hired, they filled fewer than one-third (29%) of the primary positions that advertised for someone with their specialty. Similarly, in positions with a secondary focus in professional and technical communication, PhDs in the field filled only 5 (19%) of the positions.

In all, PhDs with specialties in professional and technical communication accepted 17 (25%) of the 69 positions filled this market year. In comparison, 14 (20%) of the 69 positions were filled by individuals with PhDs in English Literature, with some coursework in professional and technical communication, and 11 (16%) were filled with PhDs in Rhetoric and Composition. Considering that some technical and professional communication programs also train doctoral students with an emphasis in rhetoric and composition, graduates from such programs accepted fewer than half of the positions with primary or secondary focus in this area (28, or 41% of the positions). This data supports anecdotal information that current doctoral programs cannot graduate sufficient doctorates to fill the market's need. However, as several search committee chairs noted, some institutions do not desire to hire specialists in the field, seeking instead to hire generalists who have some knowledge of technical or professional communication, but who also resemble and complement the literature faculty already staffing the department. Table 5 provides all the degree information for this year's hires.

Teaching Versatility

As we discussed in the job section of this article, departments varied widely from those offering only service courses in technical or professional communication to those offering all levels of courses from service courses for all majors to doctoral degrees in technical or professional communication. In all, eighteen different program variations were identified in the survey; these eighteen programs were then grouped into eight distinct categories. As Table 6 illustrates, individuals with MA or MS degrees were hired into only two of the eighteen program varieties in comparison to PhDs in professional and technical communication, who were hired into twelve, teaching technical or professional communication programs at all levels: service courses, undergraduate minor, undergraduate major, undergraduate certifi-

TABLE 5
Primary and Secondary Hires and Their Degrees, 2002–2003

| | Positions | | |
Degrees Held by Hires	*Primary*	*Secondary*	*Total*
Degree unreported	0 (0%)	1 (4%)	1 (1%)
MA/MS	1 (2%)	2 (7%)	3 (4%)
MFA	3 (7%)	5 (19%)	8 (12%)
PhD (Field unspecified)	6 (14%)	3 (11%)	9 (13%)
PhD (Other)	5 (12%)	1 (4%)	6 (8%)
PhD (English literature)	7 (17%)	7 (26%)	14 (20%)
PhD (Rhetoric/composition)	8 (19%)	3 (11%)	11 (16%)
PhD (Professional/technical communication)	12 (29%)	5 (19%)	17 (25%)
Totals	42 (100%)	27 (100%)	69 (100%)

Note.   All of the hires with MA/MS degrees, and 7 of the 8 with the MFA, filled positions at the lecturer/instructor rank.

cate, master's, master's certificate, and doctoral courses. This finding indicates that, at least in the past year's market, PhDs in technical or professional communication were best able to fit into the greatest number of program variations.

Another interesting finding of this program and course review is that PhDs in other areas (American and Cultural Studies, Linguistics, and Instructional Technology, for example) were hired into eight of the eighteen different program variations. The reason for this versatility arises from programs' needs for individuals with expertise in specific areas, such as instructional technology and cultural studies. These areas of expertise have frequently crossed over into technical or professional communication research and theory, and departments appear to be hiring specialists further to complement and nurture their students' education in these areas.

Teaching Loads

The number of courses included in hires' teaching loads varied widely from a high of twelve (four/four/four) quarters per year to a low of two (one/one) semesters per academic year. In between these extremes came a wide variety of load numbers with a two/three semester load, or five courses per academic year, being the most common. Table 7 identifies the most commonly reported teaching loads and their prevalence in our survey.

Courses

Given the diversity of these programs and the variety of course loads, it is not surprising that new hires are expected to teach a variety of courses. We discovered that most hires were expected to be able to teach courses in some combination of these four areas:

TABLE 6

Teaching Versatility and Program Variation for 69 Hires

Course or Program Varieties				Degrees Held by Hires		
	MA/MS	MFA	PhD, English Literature	PhD, Rhetoric/ Composition	PhD, Technical or Professional Communication	PhD, Other Area or Unspecified
Service only						
Service only	X	X	X	X		X
Service, undergraduate major and/or minor						
Service, minor			X	X	X	X
Service, major		X	X			X
Service, minor, major			X		X	
Service, undergraduate major and/or minor, master's						
Service, minor, major, master's		X			X	
Service, major, master's				X	X	
Service, undergraduate major or minor, master's, doctorate						
Service, minor, major, master's, doctorate			X			
Service, minor, master's, doctorate					X	
Service, major, master's, doctorate					X	
Service, master's or doctorate						
Service, master's certificate				X		
Service, doctorate						X
Undergraduate only						
Minor only						X
Major only					X	X
Minor, major				X	X	
Undergraduate major and/or minor and master's						
Minor, major, master's	X				X	
Undergraduate certificate, master's					X	
Major, master's		X	X		X	X
Undergraduate major, master's, doctorate						
Major, master's, doctorate					X	X
TOTALS	2	4	6	5	12	8

63

TABLE 7
Teaching Load Distribution for Positions Filled

Divisions	Teaching Load (Courses)	Number of Jobs	Percentage
Unreported		12	17%
Semesters	2 (1/1)	1	1%
	3 (2/1)	2	3%
	4 (2/2)	7	10%
	5 (2/3)	18	26%
	6 (3/3)	11	16%
	7 (3/4)	1	1%
	8 (4/4)	10	15%
	10 (5/5)	1	1%
Quarters	5 (2/2/1)	1	1%
	6 (2/2/2)	2	3%
	7 (3/2/2)	1	1%
	9 (3/3/3)	1	1%
	12 (4/4/4)	1	1%

1. Technical and professional communication
2. Composition and rhetoric
3. Literature
4. Other writing courses (e.g., technical journalism, creative nonfiction, creative writing)

Table 8 displays the expected teaching assignments of the hires and illustrates the various combinations of courses they might teach. It includes information on both primary and secondary filled positions and shows that new hires in these positions are most often expected to teach a combination of technical and professional communication and composition/rhetoric courses. The second most common assignment is to teach technical and professional communication courses only.

Salaries

Like teaching loads and course assignments, salaries earned by job hires varied greatly. We asked our respondents to identify their hires' salaries within a $5,000 range. Table 9 identifies the number of jobs that fell within each $5,000 range, the ranks of the individuals who accepted these jobs, their job's emphasis (either primary or secondary) in technical or professional communication, and prior education or employment histories of the hires. As in our previous analysis, we are using the categories of primary and secondary as based on the job ads.

The lowest salary range ($30,000–$35,000) was primarily populated with lecturers and teaching fellows who were recent graduates. The second range

TABLE 8
Program Varieties and Courses Hires Will Teach Within Them

Program Varieties	TPC Only	TPC and Composition/Rhetoric	TPC, Composition/ Rhetoric, and Literature	TPC and Literature	TPC, Writing Courses	Not Reported	Total
			Courses Hire Will Teach				
Service only	13	5	2	1	1	0	22
Service, undergraduate major and/or minor	1	7	1	0	2	1	12
Service, undergraduate major and/or minor, master's	4	1	1	0	0	0	6
Service, undergraduate major and/or minor, master's, doctorate	3	4	0	0	0	0	7
Service, master's and/or doctorate	0	4	0	0	0	0	4
Undergraduate major and/or minor	0	5	1	1	1	0	8
Undergraduate major and/or minor, master's	4	2	1	0	0	0	7
Undergraduate major and/or minor, master's, doctorate	1	2	0	0	0	0	3
TOTALS	26	30	6	2	4	1	69

Note. TPC = Technical and professional communication. 42 of the 69 positions filled involve undergraduate teaching only; 13 include teaching at the masters level, and 14 include doctoral courses.

($35,000–$40,000) was composed entirely of assistant professors who were either recent graduates or currently employed academics. The most common salary range for new hires was $40,000–$45,000, with 33% of the reported salaries falling within this range. These individuals were also primarily hired as tenure-track assistant professors; they all held terminal degrees but came from both graduate school and academic positions. These numbers correspond to Society for Technical Communication's Academic Salary Survey, which reports an average salary for assistant professors of $44,220 (see Sandi Harner's article in this issue, Table 1). The $45,000–$50,000 range was the most varied in rank with tenure-track assistant professors, advanced assistant professors, and an associate professor. The $50,000–$55,000 range was populated by tenure-track assistant professors with terminal degrees, and the over $55,000 range was also a mix of full, tenure-track assistant, and adjunct assistant professors, all of whom hold PhDs.

Table 9 also shows that recent graduates are more likely to fall into the lower ranges (from $30,000 to $40,000); and the data indicates that the more experience individuals have in academe, the more likely they are to command a higher salary. (This finding further supports our earlier finding that search committees place high importance on teaching and research experience and extends this finding to demonstrate that chairs and deans appear to be willing and able to compensate individuals more highly if they have these valued experiences.) Similarly, the two individuals with industry experience contracted for salaries between $45,000 and $55,000.

Individuals with primary and secondary emphases in technical or professional communication were hired in all six salary ranges, as were individuals with master's and doctoral degrees (including those without terminal degrees). What appears to distinguish these hires is not their positions' focus primarily or secondarily on technical or professional communication but on the hiring department's definition of the job as a specialist or a generalist. Individuals who accepted positions that were described with the term "generalist" tended to accept salaries in the lower salary ranges. Positions in the higher salary ranges tended to be more specifically defined, typically required expertise in electronic technologies, and often requested some administrative experience or a desire to perform administrative duties.

Other factors, not represented in Table 9, also appear to influence the hire's salary range. These factors include location of the position, program emphases, and previous retention negotiations. Programs located in East and West Coast urban centers tended to offer higher salaries, most likely because of cost-of-living expenses. Programs that had both undergraduate and graduate programs also tended to compensate hires more highly. Within these programs, it seems likely that cases have been made to separate the salaries of technical and professional communication faculty from those of literature faculty. It is also likely that more established programs have lost or nearly lost faculty members. Retention negotiations, whether they are successful or not, can increase awareness of the need for higher salaries for faculty members in this field.

TABLE 9
Salaries and Position Descriptions

Salary Ranges

	30—35K	35—40K	40—45K	45—50K	50—55K	Over 55K
Positions	11	9	21	12	5	4
Ranks	5 lecturers, 5 fellows, 1 tt assistant	1 adjunct assistant, 8 tt assistant	19 tt assistant, 2 unreported ranks	10 tt assistant, 1 advanced tt assistant, 1 associate	5 tt assistant	1 full, 2 tt assistant, 1 adjunct assistant
Emphasis	6 primary, 5 secondary	5 primary, 4 secondary	13 primary, 8 secondary	8 primary, 4 secondary	5 primary	3 primary, 1 secondary
Education	6 PhD, 5 MFA or ABD	1 unreported, 1 double master's, 7 PhDs	2 MFA, 19 PhDs	1 unreported, 1 MS, 9 PhDs	1 MFA, 4 PhDs	4 PhDs
Employment	11 graduates	6 graduate, 3 academic	14 graduates; 6 academics, 1 unreported	1 industry, 5 graduate, 6 academic	1 industry, 1 graduate, 3 academic	3 academic, 1 graduate

Note. tt = tenure track.

DISCUSSION

In the following sections, we interpret the data we have reported with the aim of identifying trends and needs regarding the market, and preparation of a faculty who will help the field develop. One broad conclusion is that there are fewer jobs for specialists than anecdote has led us to believe, and many of the jobs that presumably are for specialists involved mixed teaching assignments, frequently including composition but also sometimes including literature, creative writing, or journalism. Another broad conclusion is that too few of the jobs for specialists are being filled by graduates of the specialized degree programs. Our research confirms a significant gap between the number of available jobs and the number of qualified applicants, even though the absolute numbers may be lower this year than in the past.

Effect of the Economy

The market in 2002 was less robust than in 2000 or 2001, based on the number of jobs advertised. The obvious explanation for the softening academic job market is the economy. Many of the colleges and universities that offer courses or programs in technical communication are state supported, and all states experienced budget problems in 2002. The number of jobs canceled because of the economy (seven positions of the twenty-two unfilled) confirms that the economy influenced the shrinking of the market. The soft market for practitioners also makes academic programs less appealing, and potential new programs may be on hold until colleges can hope to place graduates.

However, it may be shortsighted to assume that the economy is the only reason for a softening academic job market. We have no hard evidence to support other conclusions, but the field should remain watchful about these possibilities: (1) The jobs that were available in 2000 and 2001 have mostly been filled; (2) programs do not plan to expand; or (3) programs that failed to hire in the past have abandoned their plans to develop programs in technical and professional communication. Although our research confirms the need for more faculty with degrees in technical communication, no one wants to get into the situation of overproduction and unemployable PhD graduates.

The Specialization Issue

It is disturbing that only 29% of jobs that named technical or professional communication as the primary area of specialization hired graduates of the doctoral programs that offer degrees in this field. Of course, faculty members prepared in English studies broadly defined can develop productive teaching and research careers in the field (many before them have done so), but they are initially handicapped by relative unfamiliarity with the field's body of knowledge and research methods. They may identify with the field and become enthusiastic participants in its work, or they may regard their teaching of technical communication as marginal to their interests. Even in the most optimal situation, the field as a whole suffers if the people who hold faculty positions are limited or delayed in the research they can do. They may spend a lot of time postgraduation developing the knowledge that a specialized degree program provides. Furthermore, if department chairs and deans are satisfied to hire people for specialist positions who hold MFAs or degrees in literature, the field will continue to struggle for academic recognition as something more than a service course that anyone who is good in "English" can teach. Work experience can compensate for academic preparation up to a point, but it is better compensation for teaching than for research. A practitioner learns methods of document development and production that create competence and credibility in the classroom, but whether work experience prepares a faculty member for research and directing student research is another question.

The lack of faculty specialists also jeopardizes programs. One search chair reported that the undergraduate program in technical communication would be canceled in a year, less because of ability to hire than because of lack of student interest, but faculty enthusiastic about a subject area are key to recruiting students. Another search chair told us that the program would shift directions from technical communication to online journalism, and this chair linked the change to inability to hire after two years of trying and to a lack of program focus. Faculty specialists can define program goals and give it focus. A third chair whose search failed spoke in frustration about the inability to develop a program that the university administration supported.

The Naming Issue

Naming this specialization continues to trouble programs. The job ads named "professional" (19%) and "technical" (18%) in about equal numbers. Many of the ads named both (23%): "professional and technical writing" or "technical and professional." "Business writing" was also frequently used (10%). A few other terms have emerged: "writing in the professions," "organizational writing," "scientific writing." Even more oblique references were found in specialists' positions, which did not call for professional, technical, business, or science communication but instead requested applications from individuals with expertise in "computer publishing" or "digital multimedia."

"Professional" in some cases seems to be an umbrella term, a little less specific and restrictive than "technical." One ad, for example, names "professional writing," followed by "e.g., technical writing." At the other extreme is a reductive sense of developing documents "in a professional manner," which suggests correctness and formatting more than the substantive work of developing knowledge.

The ads also divide on whether to name the work "writing" or "communication." The frequency with which "writing" (64%) rather than "communication" (34%) is the name suggests some redemption of this term, shunned in the 1980s as the work of the field was perceived increasingly to be visual and to occur in various media. Or its use may reflect the way the work in this field is perceived by departments that incompletely understand its range. Still, "writing" does not have to have the narrow connotation of inscribing words on the printed page. Writing by definition involves multiple sign systems, visual as well as verbal, and multiple media, digital as well as analog. To have something to say (to write) requires research, whether by interviews or testing or ethnographies, and other high-level mental skills, including judgments about the relative importance of data, ability to sort and organize, ability to make ethical judgments and apply them, and knowledge of varied audiences, including international audiences and people with visual or auditory impairments. The field would do well to respect this term and help others understand the fullness of its meaning.

We wondered, too, about the uncertainty of the field's name and its effect on the market itself. Did the differences we found in what ads requested and who departments hired reflect tightness of the job market (not enough specialists to fill all advertised positions, for example), or did this difference reflect a lack of knowledge about what specialists in the field research and teach? This question especially concerned us when one chair for an unfilled position told us that her search committee had written a position announcement that proved to be unfillable, that a dean's perception of what a person in this field could and would do was impractical. How this confusion or uncertainty affects the identity of the field is a question not answered by our survey, but, as one of Dayton and Bernhardt's participants wrote in response to their survey (see their article in this issue), we wonder about the consequences of the naming issue on field identity and marketability.

CONCLUSION

Our research confirms a significant gap between the demand for faculty prepared as specialists in technical and professional communication and the available PhDs in the field. However, the total number of faculty needed might be lower than many have assumed, unless the market in 2002–2003 was atypical. A total of sixty jobs for specialists is not a huge number and should not invite unrestrained growth at the doctoral level, but there is clearly room and a need for growth. The field needs more people prepared to teach and to develop the field's knowledge through research. Staffing academic programs with people who must learn on the job jeopardizes its growth, status, and identity. At the same time, the field needs responsible and strategic growth. We hope for ongoing monitoring of the academic job market so that people in various positions—search committees, graduate students, program directors, professional associations—can make informed decisions that affect not just individuals and individual programs, but also the field as a whole.

WORK CITED

Brumberger, Eva. "Looking to Ph.D. Students for Help with Recruiting Issues: What Factors Shape Applicants' Decisions?" *Proceedings 1999*: Science, Technology, and Communication: Program Design in the Past, Present, and Future. Council for Programs in Technical and Scientific Communication, 2000. 91–92. Accessed at www.cptsc.org October 25, 2003.

Gurak, Laura J. "Recruiting for Bachelor's Programs in Technical Communication." *Proceedings 1999*: Science, Technology, and Communication: Program Design in the Past, Present, and Future. Council for Programs in Technical and Scientific Communication, 2000. 90. Accessed at www.cptsc.org October 25, 2003.

Lay, Mary M. "Recruiting Qualified Students Into Graduate Programs in Scientific Technical Communication." *Proceedings 1999*: Science, Technology, and Communication: Program Design in the Past, Present, and Future. Council for Programs in Technical and Scientific Communication, 2000. 93. Accessed at www.cptsc.org October 25, 2003.

Rude, Carolyn. "An Agenda for Building Technical Communication from the Ground Up." *Proceedings 1999*: Science, Technology, and Communication: Program Design in the Past, Present, and Future. Council for Programs in Technical and Scientific Communication, 2000. 89. Accessed at www.cptsc.org October 25, 2003.

Carolyn Rude is Professor of English at Virginia Tech. Previously she directed the undergraduate programs in technical communication at Texas Tech University, where she participated in a number of faculty searches. She is past president of ATTW and a fellow of ATTW and of the Society fro Technical Communication. Her publications include the textbook *Technical Editing* and an edited issue of *Technical Communication Quarterly* on the discourse of public policy.

Kelli Cargile Cook is Assistant Professor of English at Utah State University. She was a job candidate in 1999–2000. Her research focuses on online technical communication and programmatic issues in undergraduate and graduate education. She is currently co-editing a collection on online technical communication and studying assessment methods in online education. Her other publications include articles on technical communication pedagogy, assessment, and computer-based classrooms.

The State of Research in Technical Communication

Ann M. Blakeslee
Eastern Michigan University

Rachel Spilka
University of Wisconsin-Milwaukee

There have been many attempts to assess the state of research in our field. This article is our attempt to both (1) synthesize recent analyses, opinions, and conclusions concerning the status of technical communication research and (2) propose an action plan aimed at redirecting our field's agenda for its research. We explore these questions: What are the recent research trends in our field? What is and is not promising about our recent approaches to research? Where do we need to go next? What are the critical components for a new agenda for our research?

The integrity and future of the field of technical communication depend on the quality and impact of our scholarship (Mirel and Spilka). However, we lack a complete perspective of the strengths and weaknesses of technical communication research. In recent history, there have been many attempts to assess the state of research in our field. Our persistent efforts in this regard underscore the need for reaching a common understanding of the questions, methods, and directions that are needed for our research.

To assist with bringing about this common understanding, we have looked at a variety of recent attempts to analyze the status of research in technical communication. We have found that, although these analyses overlap considerably, it is also often the case that they focus on different aspects of research and sometimes even result in different conclusions. This article is our attempt to both (1) synthesize these recent analyses, opinions, and conclusions concerning the status of technical communication research and (2) propose an action plan aimed at redirecting our field's agenda for its research.

Our overall goal is to provide comprehensive answers to the following questions:

- What are the recent research trends in our field? What is promising and what is not promising about recent approaches to research in the field?

• Where do we need to go next? What are the critical components for a new agenda for research in our field?

In the first part of the article, we synthesize current opinions about the quality and consistency of our research, we address the field's uses of research methodologies and the quality of our training in research methods, and we address relationships within our field (e.g., between academics and practitioners) and between our field and related fields. We also consider how research is recognized and supported within and beyond our field. As we address these issues, we consider how our recent approaches to our research have been both helpful and harmful to the integrity, vitality, and influence of our work.

In the second part of the article, we propose an action plan aimed at focusing and improving approaches to research in our field. This plan involves holding a series of forums to set goals for the future of technical communication research. These forums would occur at the major conferences in the field and would encourage input from as many academic and practitioner researchers (and trainers of researchers) as possible. The next step would be to develop guidelines and standards for technical communication research and then to establish a professional mechanism for ensuring continuous sharing, support, and information exchange regarding our research.

In developing this article, we have consolidated conclusions about the quality, status, and future of research in technical communication from the following sources:

• A survey of group facilitators at the 2003 Research Network at the annual ATTW Conference about what impressed or concerned them about participants' research-in-progress. (Facilitators included Davida Charney, Brent Faber, John R. Hayes, Karen Schriver, Stuart Selber, and Dorothy Winsor.)
• A survey of current members of the ATTW Research Committee, which Spilka chairs, and of current ATTW officers; of strengths, weaknesses, and needs of technical communication research. (Committee members include Ann Blakeslee, Davida Charney, Dave Clark, Brent Faber, Roger Grice, Karen Schriver, Rachel Spilka, Clay Spinuzzi, and Mark Zachry. ATTW officers surveyed include Sam Dragga, Carolyn Rude, Stuart Selber, and Dorothy Winsor.)
• A survey of faculty representatives from schools with established PhD programs in technical communication. (Programs surveyed include Iowa State University, Dorothy Winsor; Michigan Technological University, Bob Johnson; New Mexico State University, Barry Thatcher; Penn State University, Jack Selzer; Purdue University, Graham Smart; Rensselaer Polytechnic Institute, Bill Hart-Davidson; and Texas Tech University, Carolyn Rude.)
• Position papers and discussions at the 2000 Milwaukee Symposium. (The Symposium took place at the University of Wisconsin-Milwaukee. Its aim was to identify problems of the field and desirable or necessary directions for its future.

Papers and discussions focused on recent trends in the field and major changes that seem necessary. Barbara Mirel [Univerity of Michigan] and Karen Schriver [KSA Consulting] were session moderators. Other participants included Stephen Bernhardt [University of Delaware]; Russell Borland [retired from Microsoft]; Deborah Bosley [University of North Carolina-Charlotte]; Stan Dicks [North Carolina State University]; Roger Grice [IBM and Renssalaer Polytechic Institute]; Kathy Harmamundanis [Compaq Computing]; Johndan Johnson-Eilola [Clarkson University]; Susan Jones [Massachusetts Institute of Technology]; Jimmie Killingsworth [Texas A&M University]; Leslie Olsen [University of Michigan]; Jim Palmer [NCI]; Judy Ramey [University of Washington]; Mary Beth Raven [Iris Associates]; Stephanie Rosenbaum [Tech-Ed, Inc.]; Rachel Spilka [University of Wisconsin-Milwaukee]; and Elizabeth Tebeaux [Texas A&M University].)

• Selected chapters from the 2003 anthology, *Reshaping Technical Communication: New Directions and Challenges for the 21st Century* (Erlbaum, 2002), edited by Barbara Mirel and Rachel Spilka.

• Selected chapters in Blakeslee's forthcoming textbook with Cathy Fleischer, *Becoming a Writing Researcher* (Lawrence Erlbaum Associates, Inc.).

• Three recent conference presentations: Blakeslee's presentation at the 2002 Conference on College Composition and Communication (CCCC) on the state of the graduate research methods class, Schriver's presentation at the 1998 ATTW Conference on concerns about recent approaches to technical communication research by academics, and Doheny-Farina's 2000 keynote address at the annual ATTW Conference on the state of research reporting in the field.

• A literature review of technical communication research discussions in *Technical Communication*, *Technical Communication Quarterly*, and the *Journal of Business and Technical Communication* during the past eight years.

We hope that our synthesis of recent accolades and critiques of technical communication research, along with our proposed agenda, will help in two vital ways. First, we hope that this article will inspire scholars to look internally, to examine how they personally—as individual researchers and teachers of research—can contribute to sustaining the strengths and overcoming the weaknesses of research in our field. Second, for even greater impact, we hope that this article will inspire communities of scholars in our field to work as a collective whole toward setting goals and suggesting productive agendas for our research. Ideally, everyone who conducts, teaches, or gains from technical communication research will be highly motivated to participate actively in upcoming discussions led and coordinated by our major professional organizations and directed toward substantive, large-scale, and long-lasting change. Action at both levels—the personal and the collective— will be key, we are convinced, to ensuring that our future research is recognized for its integrity, vitality, and impact.

QUALITY AND CONSISTENCY OF RESEARCH

Quality seldom is easy to define. In fact, according to almost all researchers we surveyed, what counts as quality research in technical communication is not well articulated. Many also envision the need for a consistent, thoughtful, and well- articulated approach to ensuring quality in our research. Their comments in this regard covered many different aspects of our research, including how we approach it, the breadth and depth of our research, and the amount of research we are conducting.

Our Approach to Research Needs to Be Consistent, Systematic, and Thorough

The researchers we surveyed, along with the participants in the Milwaukee Symposium, agree that a main goal for our research should be to strengthen how we approach it. We need approaches, they stress, that are well-planned, coherent, and systematic, and that lend rigor and validity to our research.

Many researchers we surveyed problematized how we approach research in technical communication. Faber, for example, expresses concern that some researchers are jumping too quickly to theory before collecting, interpreting, or fully understanding data. He believes that research in our field is too often predetermined to fulfill theoretical models rather than being used to challenge or build onto such models. Selber agrees and urges that we let theory inform rather than direct our research. According to Selber, we should let our research dictate the theoretical lenses we use instead of taking top-down approaches that simply satisfy theory.

Researchers are also concerned about how systematic we are in approaching our research. Charney, for example, is concerned that we may be underestimating the planning, decision making, and scholarly review that is needed in research. Winsor and Selber concur that we sometimes approach research, especially parts of the research process like data analysis, without sufficient rigor.

We Need to Build a Coherent Body of Knowledge

Although our surveys of researchers suggest a lack of consistency and thoroughness in how we approach research, they also expose some strengths, especially in relation to the coverage of our research. For example, most researchers surveyed agree that we cover many topics quite well (e.g., the history of technical communication, technology, and computer-mediated communication). By "well," these researchers mean that the topic has been researched a fair amount, and that the field is advancing in its understanding of it. However, they are also concerned that too few people are working on complementary research questions leading to a coherent body of knowledge on certain topics. Zachry and Rude, for example, both point out that too much research in our field is driven by individual interests and inclinations rather than by some overarching initiative. One reason for this is that we lack a shared agenda for our research, a problem that Zachry attributes to the paucity of funding opportunities in our field (we address this problem again in the recognition and support section).

Others we surveyed note a similar problem of researchers not building on each other's work. They perceive problems with how authors summarize existing research in the field and with how well they build on past research. Winsor, for example, notes that research studies often do not relate well to each other, and Charney argues that overlap among projects in technical communication is insufficient for either building on or challenging published work. Rude and Selber recommend that we start paying attention to whether we are asking the same questions or retracing questions in ways that prevent us from moving forward. A discussion on the ATTW listserv (ATTW-L) also addressed this issue, suggesting that technical communicators have not done enough to summarize existing research or to build on past studies (April 23—28, 2003, "The Reach of Research in Our Field;" available through the archives). Because learning from the past is a vital means of moving forward and maturing, the field could benefit from a greater effort to identify what is valuable in existing research and to either replicate or respond to studies already completed.

We Need Agreement About Key Questions for the Field

The researchers we surveyed went further than simply suggesting the need to replicate existing work. They all concurred that we need to agree upon specific, broad questions that we consider important for our field to explore, and we need to articulate these questions in a clearer and more focused manner. Many of the researchers surveyed sense that we are having difficulty as a field articulating research questions that are appropriate and useful. Charney, for example, notes that researchers sometimes approach their work with a solid sense of their methods or of an area of investigation, but not with particular questions in mind: They assume that the projects will take shape during field work or when they finally analyze the data.

Spinuzzi proposes that we develop clear research questions that are specific to our field, and Selber and Rude suggest that the field ask itself what it most needs and what will benefit it. Rude also suggests defining broad research questions that the field can then work to answer. She argues that we need a clearer sense of what the field needs to know, along with a more aggressive approach to inquiry. Charney extends these suggestions:

> Leaders in the field need to identify a few specific major questions to spur related work from a wide range of scholars. The questions should be explicitly posed in a

way that encourages them as a nationwide focus for graduate seminars, dissertations, and eventually conference panels and special issues. (personal communication)

According to the researchers surveyed and the participants of the Milwaukee Symposium, one particular area where we need to define more and better questions is in relation to industry. We found universal agreement that the needs of industry should have at least some influence on the questions we articulate. Participants in the Milwaukee Symposium, in particular, express a strong desire for researchers to investigate more research problems that industry considers important. They also stress that this research needs to lead, eventually, to guidelines and best practices for the field. They identified many such questions in the categories of user and task analysis, communication in context, design, usability, information access and retrieval, international communication, and project management (for a full list of their questions, see the Appendix of Mirel and Spilka). Of course, one concern with such efforts is that we not limit ourselves or stifle our creativity. In a field as dynamic as ours, we need always to be receptive to new issues and questions.

In thinking about the kinds of research questions we should be asking, we also need to consider the question of what counts, or does not count, as legitimate research in the field. A few in our survey, for example, raise questions about the legitimacy of research on teaching. However, many others argue that teaching is a complex process whose methods and results should be tested just like those of other complex processes, especially as new theories and technologies offer opportunities for innovation in teaching. Without research, how do we know, for example, what works in online teaching in writing? Or how do we know what the results are of service-learning projects? A few graduate programs actively discourage such research, but the survey of ATTW members that Dayton and Bernhardt conducted, and that is reported in this issue, reveals high interest in articles on teaching with technology and online teaching.

We Need Less Thinking and Talking About Research and More Doing It

Finally, in commenting on the quality and consistency of our research, many of those surveyed agree that we need to do more actual research and less analyzing of our research. The shared perception is that too much of our scholarship (and publication) is based on introspection and philosophizing. While a certain amount of introspection is useful, any field needs to be careful of doing it at the expense of actual research. As Winsor and Smart point out, we tend to do more talking about research than actually doing it. Similarly, participants in the Milwaukee Symposium stress that we also need to carry out more primary research, especially longitudinal research. There is certainly some validity to a certain amount of introspection, as we hope this article exemplifies, but as a field, we need to be careful to not go too far in this direction.

METHODOLOGY AND TRAINING IN METHODOLOGY

We Benefit From But Should Be Careful About Borrowing Methods From Other Fields

One reason we may struggle with our approaches to research is our common practice of borrowing methodology from other fields. Our surveys revealed that some of us use methods borrowed from the social sciences, including quasi-experimental and other quantitative methods as well as qualitative methods (e.g., ethnographies and case studies). Still others use rhetorical and discourse analytic techniques borrowed from rhetoric and linguistics. Our methodologies are thus quite varied. As Rude indicates, technical communicators preparing to do research have both multiple methods available to them and multiple sites of practice, including industrial, scientific, and social organizations.

Most researchers surveyed characterize this variability in methodology as a strength. However, some also note its liabilities. Spinuzzi, for example, points out that we have not consistently explored the methodological implications of mixing and matching approaches. Dragga also cautions against ill-informed borrowing and adaptation of methods. Johnson, in a *TCQ* article addressing interdisciplinary methods, also addresses the risks of using methods from outside the field. Johnson argues that "technical communication must assume the…responsibility of understanding the ideologies, contexts, values, and histories of those disciplines from which we borrow before we begin using their methods and research findings" (75).

To address these risks sensibly, researchers need to ask themselves how carefully they consider the methods they use and how appropriate these methods are for the questions they are asking. Rude similarly argues that we need to define our methods better. She notes that our current collections offer just chapters rather than substantial and sustained theorizing, along with procedures, that help researchers know what to do and how to do it. (Blakeslee and Fleiseher's forthcoming research textbook was conceptualized as a response to this problem.) Dragga agrees and calls for additional work in our field to explain and debate the merits of our methods.

While acknowledging the potential drawbacks of using borrowed methods, many researchers stress that this borrowing is mostly an asset. Winsor argues, for example, that our sophistication with methodology is increasing. She adds that, overall, we tend to make good use of multidisciplinary methods and theories. Others surveyed also see our uses of methods becoming increasingly sophisticated and view them as being quite flexible and even innovative. The consensus, therefore, seems to be that our field benefits from the wide variety of methods available to it and from the discussions and explorations of these methods that have occurred in other fields. As Winsor points out, we have not partitioned ourselves off from that which is useful to us; the trick is to be sure that we adapt these methods appropriately for our own uses.

We Need to Be Aware of Which Research Methods We Use and How We Use Them

While the perception is that our methodological pluralism is more of a strength than a weakness in our field, the ways in which we sometimes use various research methods may be problematic. Some of the researchers we surveyed suggest that a tendency to do what is most comfortable or most convenient for us may limit our research. Charney questions whether we have a good enough sense of which methods are helpful for which questions, and she proposes that we strive to do a better job, overall, of matching methods with questions. Others argue that we do too much of certain kinds of research and too little of other kinds. Rude, for example, has observed that we may be doing too much analyzing and critiquing of texts when maybe other approaches could or should be chosen instead. Like Charney, she addresses the need to match our methods to our questions.

Perhaps the most fruitful place to develop a sense of which methods researchers are using and how they are using them is at the graduate level. What are our up-and-coming scholars doing? In our survey of graduate programs in technical communication, we found a strong correlation between the methods that students in these programs use and those that their faculty use. For example, at Pennsylvania State University, faculty told us that what they do best is historical and rhetorical work. Many of their students, therefore, are doing that type of scholarship. At Iowa State University, most students are choosing to do qualitative research, primarily interviews and observations, although about 20% are doing statistical work and another 20% are doing text analysis. These choices again reflect faculty interests. Dissertation topics at Iowa State also illustrate the influence of faculty interests: Students recently have been working on projects in visual rhetoric, power in the workplace and classroom, WAC, activity theory, and learning communities, all interests that can be connected to particular faculty.

At Michigan Technological University, students tend to do textual/rhetorical analyses grounded in rhetorical or cultural theory. Common topics include visuals, narrative theory, feminist theory, environmental policy, design, pedagogy, and usability. Again, both the methods and the topics correspond closely to faculty interests. At Texas Tech University, students use qualitative and quantitative methods and rhetorical analysis, and at Purdue University, a fair amount of the doctoral research involves studies of the discourse of listservs and Web sites, which again reflects faculty interests and strengths. We also surveyed Rensselaer

Polytechnic Institute and New Mexico State University with similar results. At New Mexico State, we observed a geographical influence with students showing interest in intercultural and border rhetoric. RPI stresses its interdisciplinarity and claims that, as a result of this feature, their students are exposed to a wider variety of methods.

We Need More Consistent, Systematic, and Extensive Training in Methods

Despite the varied and interesting projects we found graduate students doing, we heard in our research numerous concerns about the state of our training in methodology. Participants at the Milwaukee Symposium bemoaned the inconsistency of what we cover and of what students learn in our research courses. According to participants, we need, as a field, to identify and agree on basic research competencies and then to ensure that we train students in those competencies in our programs. One potential problem in this regard is that the faculty preparing doctoral students may not themselves be well prepared in empirical methods. They may be cautious, therefore, about teaching or directing research when they themselves lack expertise in the methods. A related issue that researchers addressed is how we lack both teaching materials and courses in technical communication research methods. Dragga notes how we often farm out such training to other areas. He recommends that we develop more of our own courses in research methods, along with more textbooks.

Because of credit hour restrictions, many graduate programs in technical communication are constrained in the number of methods courses they can even offer. Several researchers indicated that their programs offer only one course, which, as Winsor notes, is not sufficient for training students to be researchers. One of us, Blakeslee, has researched how graduate students develop their identities as writing researchers. In a 2002 paper delivered at CCCC, she argued that we need to pay greater attention to the research training of our students and attend more to their experiences as they develop as researchers. We also need to rethink our pedagogical approaches to our research courses, focusing more on the process of research and on students' experiences as they develop their identities as researchers (see also Rose and McClafferty, and Vol. 30 of the journal, *Educational Research*). Blakeslee argued, as well, for remembering that our students will conduct research in a variety of settings and for a variety of purposes. While some may end up doing primarily academic research, others may end up doing applied research as practitioners in professional settings.

Because of its own interdisciplinary emphasis, RPI offers a useful model for how to train students in research methods. At RPI, students can take ethnography and qualitative methods from cultural anthropologists in their own department or in the department of science and technology studies. They also all receive an introduction to social scientific communication methods with an emphasis on experimental and quasi-experimental design and statistics. They learn, as well, various kinds of hybrid methods combining qualitative and quantitative data gathering and analysis (e.g., verbal data analysis, usability testing, and user-centered design). While not every program is equipped to offer this much variety, and probably most are not, programs should at least endeavor to offer depth in some methods while integrating and supporting methodological training, more generally, throughout the curriculum.

RELATIONSHIPS WITH PRACTITIONERS
AND WITH OTHER DISCIPLINES

Our field is clearly interdependent and interdisciplinary. Within technical communication, the work of academics needs to connect closely with the work of practitioners. While academics need to keep learning from practitioners about effective ways to prepare students for the complex kinds of thinking and writing that they will encounter in workplace contexts, practitioners can benefit from learning about academic research findings and then adapting them, with caution and rhetorical sensitivity, to their own organizations. Similarly, technical communicators find it extremely helpful to borrow routinely from related fields. Just as technical communication academics almost universally value and use theories and methods from related disciplines, practitioners in our field find it natural and useful to emulate approaches to practice from related occupations. Many academics and practitioners discover technical communication only after devoting time in related careers such as human factors, human resources, public relations, and business management. Some even migrate to those other, related careers after short or long stints in technical communication (Anschuetz and Rosenbaum). Because technical communicators do not, by definition, work in isolation, the field needs to ensure the vitality and quality of (1) the academic-practitioner relationship within technical communication and (2) the field's relationship with other, related disciplines.

We Need to Improve the Academic-Practitioner
Relationship Within Our Field

Unfortunately, the relationship between academia and industry in technical communication has always been somewhat strained. Not surprisingly, the discourse addressing this relationship has been substantial (see, for example, Mirel and Spilka's anthology, in which the entire first half of the volume is devoted to describing this problem and possible solutions for it). Although it is always risky to oversimplify complex relationships, such as those between academics and practitioners in technical communication, it is safe to conclude that many practitioners in industry lament that academic research often lacks relevance to workplace realities and is of little value to them. It is also safe to conclude that many academics are concerned about the apparent dearth of a theoretical foundation for workplace practice, about the overly practical nature of practitioners' work and interests, and about the perception that practitioners simply do not welcome academic input or involvement in their work.

The problematic relationship between academics and practitioners certainly does seem significant. At times, it seems as though the divide between us will never be bridged. At best, improving this relationship is likely to require solutions that are farther reaching and more powerful than those tried to date (see Mirel and Spilka, Part I, Introduction). Zachry suggests looking to other fields that also deal with splits between academics and practitioners. He says we could possibly learn strategies for our own field by studying how others have tried to join academic and practitioner interests.

Academics need to improve efforts to show the relevance of their research to industry. On a positive note, many academics see the considerable relevance of academic research to workplace practice. Unfortunately, as Blakeslee points out, academics often fall short in their failure to make a substantive connection with practitioners (that is, a connection beyond sharing research findings with the limited audience of professionals at the original research site). As Schriver (ATTW conference) and others in the Mirel and Spilka anthology point out, much academic research never finds its way to practitioner audiences. One reason for this is that most academic scholars in our field continue to publish in journals and present at conferences targeted at academics. Rude argues that our research has been most useful and influential to industry through the publication of books such as those by Schriver on document design and by Hackos on project management, which have sold well to practitioners. If more technical communicators could produce similar types of publications that synthesize research—both in our own and in related fields—in a way that practitioners find accessible and relevant, we could make significant strides in showing practitioners the potential value of academic research to their work. (See the "Recognition" section for further discussion of this recommendation.)

We need to provide more opportunities for academics to interact with practitioners and to collaborate on joint projects. Participants at the Milwaukee Symposium suggest that the field make more forums available (like the Milwaukee Symposium) for exchange and collaboration between practitioners and academics. Toward that goal, they advocate more service-learning and actual client projects for academic courses in technical communication, an argument also made by Blakeslee in the *Reshaping* volume. They also propose the following:

- More community projects that combine the expertise of both academics/students and practitioners

- The establishment of academic boards of industry representatives who, among other services, can inform academic institutions about ways to make instruction and curriculum designs more relevant and useful to the needs of industry
- Greater priority for internship courses in academic programs
- More consulting opportunities for students and faculty in industry settings
- The development of centers that involve academia-industry collaborations

Another recommendation is that more symposia take place so that academics and practitioners can have more opportunities to share ideas and resources. Participants in the Milwaukee Symposium also recommend the development of joint academic/practitioner research projects. The consensus is that faculty need training in how to identify industry partners for research projects and in how to work productively (and with minimal tension) with industry partners in all phases of research.

We need to overcome differences between academics and practitioners to facilitate collaboration between the two worlds. Participants at the Milwaukee Symposium also emphasize the need to overcome discourse differences between academics and practitioners (a problem also described by Dicks in Mirel and Spilka's volume). Contributing to academic/practitioner tensions are differences in how academics and practitioners describe the need for research projects, their design and methodology, the obstacles that exist to research and their solutions, methods for data analysis, and so on. When academics and practitioners collaborate on research projects, they need a common vocabulary. Otherwise, differences in perspective are likely to create tensions.

Participants at the Milwaukee Symposium agree that academics and practitioners alike need to identify and overcome institutional barriers that work against collaboration. For example, in many universities, faculty members have no reward system for collaborating with practitioners, and they find that many of their colleagues devalue (and even scorn) this type of research. Similarly, many organizations are suspicious about research collaborations with academics. Practitioners may see little value in academic contributions. Others are concerned about possible proprietary problems, loss of productivity, and exposure of industry problems. These concerns often override any value they might see in academic/practitioner collaborations. One industry participant at the Milwaukee Symposium confided that her superior had been reluctant to have her attend the Symposium. In his view, her two days away would result in a great economic loss that would override any benefit from her contributions to a national discussion about the status and future of technical communication. He had trouble understanding how any insights that she would gain from the Symposium would benefit his organization: He had no interest in her contributions to a cause that, in his mind, had little to do with his company goals. An important first step for academics and practitioners who wish to collaborate, and even for those who just want to improve that relationship, will be to educate others about the value of such efforts, both for the institution and for the field.

We Also Need to Improve Our Relationship with Other Fields

Given the extent to which we draw on and borrow from other fields, our relationships with those other fields are also critical for our continued vitality and growth. Unfortunately, our relationships with some disciplines, particularly those with which we are closely aligned, have also become increasingly strained over the past decade. This may be a result of increasing competition with the other fields about who "owns" or "should own" information and Web design (Schriver). We hope that the field will soon collectively address this concern.

We need to encourage continued collegiality with those in related fields.   Charney cautions us from separating ourselves too much from other fields, including rhetoric and composition, from which many of us developed the theoretical and pedagogical foundations for our work. According to Charney, there is danger in repeating mistakes of the past by isolating ourselves completely from issues in writing instruction, theory, and scholarship. Because we share many concerns with these colleagues, we stand to experience greater gains than losses from maintaining our connections with them.

We need to give top priority to overcoming differences and discovering commonalities between our field and related fields, especially other technology fields.   Participants at the Milwaukee Symposium recommend that academics and practitioners alike learn the discourses of related fields and use these discourses more effectively to express the complexity behind what we do and produce. This group also urges us to display greater pride in our work and improve our marketing of it, thus improving our image. Again, if we wish to extend our visibility and impact beyond our own field, we need to begin communicating our research beyond our own journals and conferences. We also need to ensure that our journals are indexed in the major online databases. The participants in the Milwaukee Symposium also recommend that we identify and then work through tensions about "ownership" and work to acquire greater institutional power so that we can engage in large-scale collaborations with related fields.

We need to generate more opportunities for cross-disciplinary collaborations.   From an ATTW listserv discussion, we found that a growing number of academics are participating already in such projects (April 23–28, 2003, "The Reach of Research in Our Field"). Many researchers described successful ways in which they became involved with cross-disciplinary research. This is a promising start. As Barchilon suggests, it could help to survey faculty already collaborating with members of other fields about how they succeeded in starting up and continuing those projects (April 24, 2003, "The Reach of Research in Our Field"). Another possible step would be to propose that professional organizations and societies in related fields encourage researchers to collaborate on projects, perhaps through a competition for research grants for those types of projects. On a larger scale, Zachry proposes the development of a professional infrastructure in which our professional organizations would take the lead in finding interdisciplinary collaborators to partner with technical communication researchers. This type of infrastructure would also lend financial and other support, along with increased visibility, to such efforts.

Recognition, Support, and Resources

The strength of relationships between academics and practitioners, between practitioners and others at work sites, and between technical communicators and constituents in other fields is, as suggested in the previous section, in large part a matter of visibility. Our field needs to make its work visible, and it must get it recognized sufficiently, particularly by practitioners and by other fields for which it has relevance. The Milwaukee Symposium, for example, identifies as two major goals for the field to elevate our status and to broaden our influence. We struggle, especially, with finding sufficient resources for our work. We need, therefore, to identify ways to increase the visibility and influence of our work, as well as to increase the levels of support and funding for it.

We need to disseminate our research findings broadly and increase readership of our journals.   To enhance the visibility of our work, we need to present it in ways that are accessible, relevant, and meaningful to readers. We also need to present it in venues that they are likely to consult. Participants in the Milwaukee Symposium suggest that researchers deliberately seek out new forums for reporting their work, including local, community, and public forums. Both Doheny-Farina, in his keynote address at the 2000 ATTW Conference, and Schriver, in her 1998 presentation at this conference, made similar arguments, and both offered concrete examples of how we might accomplish this.

Because most technical communication research is published in academic journals, and because academics concerned with tenure need to consider the kinds of journals they publish in, they may be constrained in considering alternative venues for their work, especially nonacademic ones. Not all academic researchers have the flexibility of publishing in just any forum. Even so, academics need to work more diligently to help their colleagues understand why they might need to do this and why it is of value. They also need to consider options such as publishing first in a forum that "counts" for them and then adapting their work for presentation in different forums that reach their wider audiences (Schriver, ATTW Conference).

For practitioner researchers, a serious and frequent constraint is that of proprietary issues. At the Milwaukee Symposium, practitioner participants said that they often were discouraged or forbidden from sharing their research because of corporate regulations. Companies are often concerned about sharing problems and research findings with external audiences. A goal for the field should be to identify more effective ways to cope with proprietary issues, because they impede efforts to disseminate research to broader audiences.

Academics need to affiliate with organizations in related fields and increase their involvement both in those organizations and on key committees both internal and external to their institutions. The issue of visibility is connected not only to how we disseminate our work; it is also a matter of how we present ourselves and participate in the field. How many of us, for example, have served on our campus's Institutional Review Boards? To what extent have members of our field been involved with major research foundations? In the ATTW-list discussion cited earlier, Bernhardt advocates joining associations and committees with people from industry (April 26, 2003, "The Reach of Research in Our Field"). He offers the Drug Information Association, to which he belongs, as an example. Grabill also talks about becoming involved in organizations and influencing practice. There are numerous ways in which we can make ourselves more visible outside of our immediate departments and professional associations, and we should actively pursue opportunities that can enhance both our own visibility and that of the profession.

Practitioners also need to find ways to increase leverage within their domains to produce quality outcomes. Milwaukee Symposium participants identify as another central concern the limited impact of practitioner research due to institutional barriers that constrain their influence to a limited area within their organizations. To increase the sphere of practitioner influence, the participants propose educating practitioners about how to move through and elevate their status in organizations. They also propose educating them about how to assume new roles that can bring about significant change in the workplace. For example, practitioners need to learn how to recognize what they need to know and, if they do not know it, how to identify the "right people" to work with. In addition, practitioners would benefit from participating earlier in planning and decision making, and from learning how to scope out the politics of their organizations. Of course, academics also need to educate students about analyzing and then handling these complexities.

We need both internal and external support and funding.   Finally, a discussion of recognition would be incomplete without a discussion of support, whether it be from internal or external sources. Winsor expresses concern that too many researchers in technical communication never collect data again after completing their dissertations. Why is this? Faber suggests that our field, as a whole, suffers from a lack of fundable research. He also notes, along with Dragga, how we lack presence or recognition with major research foundations such as the National Science Foundation, the National Endowment for the Humanities, and the National Institute of Health. These researchers, along with the participants in the Milwaukee Symposium, argue that we should explore ways to make our research more interesting and appealing to funders. In comparison with many fields, our research is inexpensive. Still, we would benefit from greater opportunities to obtain support that would provide us with released time and with other needed resources. More funding would also facilitate both disciplinary and interdisciplinary collaborations for our research, and it would support the sort of sustained research initiatives that many of the individuals we surveyed believe our field needs.

Participants in the Milwaukee Symposium offer four solutions to the problem of outside funding. The first is to educate funding agencies about the value and nature of our research. Second is partnering with others on projects. Third is to develop guidelines for how to win research grants. And fourth is to create online research directories. One outcome of the Milwaukee Symposium was the Research Repository (currently under development by the ATTW Research Committee), which is designed to match researchers in both academia and industry with research funders and to provide information about how to obtain research support. The Society for Technical Communication (STC) Research Committee has developed a similar online service. Such efforts will go far in providing researchers with the financial support they need to do their work and to do it well.

Sources of internal support are also an important component of research. Such support includes university support to develop grant proposals, released time to carry out research, and other internal awards that provide the time and resources researchers need to carry out their work. While most of our institutions require that we do research, they sometimes give us mixed messages with the teaching loads, committee assignments, and other responsibilities they assign us. In fact, participants in the Milwaukee Symposium single out significant institutional constraints as one of the main factors impeding research in our field. They cite issues of time, support, and also incentives, and recommend that institutions make it more valuable and worthwhile for individuals to do research.

Internal support can also come from our own professional organizations. As a discipline, we need an infrastructure that supports research in our field. And while we currently have grants through STC and the Council for Programs on Technical and Scientific Communication (CPTSC), little else is available to provide either monetary support or other kinds of resources that researchers need. Zachry argues that we need our professional organizations to take the lead in identifying categories of research questions and in helping interested researchers find grant opportunities and interdisciplinary collaborators. The call on the ATTW list for a survey of individuals doing cross-disciplinary collaborations offers a good example of something our professional organizations could take a lead in doing (Barchilon, April 24, 2003, "The Reach of Research in Our Field"). The recommendations of the Milwaukee Symposium are also ones these organizations could pursue.

PROPOSED PLAN OF ACTION FOR OUR FIELD

We have identified in this article many strengths we have in regard to our research, but we have also outlined some of the challenges we face. In doing so, our hope is not to discourage members of the field from doing research but to encourage and urge all of us, whether academics or practitioners, to identify and work toward solutions. Therefore, we propose that, as a field, we move forward assertively to initiate and sustain a plan of action for ensuring the quality, vitality, and impact of our research and scholarship. In particular, we propose the following plan of action for our field and encourage our major organizations, including ATTW, CPTSC, and STC, to take the following steps within the next few calendar years.

Step One: Expand Our Problem Definition and Vision

We propose that the field coordinate a year-long series of forums, similar to the Milwaukee Symposium, in which members of the field will have an opportunity to contribute their insights regarding the following:

- The quality of research in the field
- The quality of research training in the field
- Relationships between academics and practitioners and between technical communication and related fields
- Issues of research support, status, and visibility

The primary aim of these forums will be to expand our understanding of why research in our field sometimes has little impact. A secondary aim will be to use input from a variety of academics and practitioners to formulate goals for the field and a vision of how research can add to the field's vitality and influence.

Specifics.   The ATTW Research Committee will initiate this effort in the Research Network to be held at the March 2004 Annual ATTW Conference; this will be an opportunity to collect input from many academics in the field. Further discussions, also sponsored by the ATTW Research Committee, will take place throughout the 2004–2005 academic year at the annual CPTSC and STC conferences. These sessions will be designed to collect input from administrators of academic programs as well as from industry leaders and practitioners in the field.

Step Two: Develop Concrete Solutions

We propose that between Summer, 2005 and Summer, 2006, the ATTW Research Committee summarize input to date and then sponsor several two-day retreats of the ATTW Research Committee, STC Research Committee, and a small set of CPTSC members who administer graduate programs in technical communication and also have experience teaching methods courses in those programs. At these retreats, participants will develop the following outcomes:

- A set of guidelines and standards for research in the field
- A set of research questions that will be presented as most important for the field to investigate in the twenty-first century
- A plan for disseminating the set of guidelines and standards and the set of research questions to the field
- A set of time-specific, realistic solutions for providing incentives and opportunities for academics and practitioners to follow the guidelines and standards and to investigate the set of research questions (most likely, these solutions will need to be implemented through the leadership and support of individual professional organizations or, ideally, through the support of a collaborative consortium of professional organizations in technical communication and related fields)

Step Three: Implement and Evaluate

Although this stage will evolve and become specific over the next few years, we propose that everyone involved with this effort implement solutions between Summer, 2006 and Summer, 2008 and then reconvene to plan ways to evaluate their effectiveness. Most likely, another series of surveys, focus groups, and forums will be useful toward achieving that goal.

CLOSING THOUGHTS

The plan we present is ambitious but achievable. We also believe that it is essential to enhancing and establishing the value of our research. The responses we received from the researchers surveyed certainly reveal confidence and pride in the research we have done in our field, but they also drive home the point that we could still do better. There are many ways to improve, but of greatest concern here is that our approaches to research could be more unified and systematic, our questions could be better targeted and more connected, our training in and use of research methods could be more rigorous, and the dissemination and impact of our research results could be more effective and far-reaching. Further, we need to make the case, both inside and outside of our institutions and workplaces, for sufficient support for our research. In working to achieve these goals, we have numerous strengths on which to build, and we certainly should concentrate as much on those strengths as we do on the weaknesses we need to improve and the constraints we need to overcome.

WORKS CITED

Anschuetz, Lori, and Stephanie Rosenbaum. "Expanding Roles for Technical Communicators." *Reshaping Technical Communication: New Directions and Challenges for the 21st Century.* Ed. Barbara Mirel and Rachel Spilka. Mahwah, NJ: Lawrence Erlbaum Associates, Inc., 2002. 149–63.

ATTW-L Archives. lyris.acs.ttu.edu/cgi-bin/lyris.pl?enter=attw-l.

Blakeslee, Ann. "Making Researchers: Rethinking the Graduate Research Methods Class." Annual Convention of the Conference on College Composition and Communication, March 2002.

Blakeslee, Ann. "Researching a Common Ground: Exploring the Space Where Academic and Workplace Cultures Meet." *Reshaping Technical Communication: New Directions and Challenges for the 21st Century.* Ed. Barbara Mirel and Rachel Spilke. Mahwah, NJ: Lawrence Erlbaum Associates, Inc. 2002, 41–55.

Blakeslee, Ann, and Cathy Fleischer. *Becoming a Writing Researcher.* Mahwah, NJ: Lawrence Erlbaum Associates, Inc., forthcoming.

Dicks, R. Stanley. "Cultural Impediments to Understanding: Are They Surmountable?" *Reshaping Technical Communication: New Directions and Challenges for the 21st Century.* Ed. Barbara Mirel and Rachel Spilka. Mahwah, NJ: Lawrence Erlbaum Associates, Inc., 2002. 13–25.

Doheny-Farina, Stephen. "Writing Without Discipline." TechComm 2000, 3rd Annual ATTW Conference, April 2000.

Grabill, Jeffrey T. "Shaping Local HIV/AIDS Services Policy through Activist Research: The Problem of Client Involvement." *Technical Communication Quarterly* 9 (2000): 29–50.

Hackos, JoAnne T. *Managing Your Documentation Projects.* New York: John Wiley & Sons, 1994.

Johnson, Robert R. "Complicating Technology: Interdisciplinary Method, the Burden of Comprehension, and the Ethical Space of the Technical Communicator." *Technical Communication Quarterly* 7 (1998): 75–98.

Mirel, Barbara, and Rachel Spilka. *Reshaping Technical Communication: New Directions and Challenges for the 21st Century.* Mahwah, NJ: Lawrence Erlbaum Associates, Inc., 2002.

Rose, Mike, and Karen A. McClafferty. "A Call for the Teaching of Writing in Graduate Education. *Educational Researcher* 30 (2001): 27–33.

Schriver, Karen. *Dynamics in Document Design.* New York: John Wiley & Sons, 1997.

Schriver, Karen A. "Looking Back, Looking Forward: Reflecting on the Changing Face of Research in Technical Communication." Presented at the Annual Meeting of the Association of Teachers of Technical Writing, Chicago, IL, April 1998.

Schriver, Karen. "Taking Our Stakeholders Seriously: Re-Imagining the Dissemination of Research in Information Design." *Reshaping Technical Communication: New Directions and Challenges for the 21st Century.* Ed. Barbara Mirel and Rachel Spilka. Mahwah, NJ: Lawrence Erlbaum Associates, Inc., 2002. 111–33.

Ann Blakeslee is Professor of English and Director of Undergraduate Studies at Eastern Michigan University. In her scholarship, Ann has focused on genre theory, audience, the rhetorical practices of scientists, and strategies for learning professional and academic genres. She is author of *Interacting with Audiences: Social Influences on the Production of Scientific Writing* (Lawrence Erlbaum Associates, Inc.). She is also Treasurer and research committee member of the Association of Teachers of Technical Writing (ATTW) and Vice President of the Council on Programs in Technical and Scientific Communication (CPTSC).

Rachel Spilka is Associate Professor of English and coordinator of the Graduate Program in Professional Writing at the University of Wisconsin-Milwaukee. Rachel is editor of *Writing in the Workplace: New Research Perspectives* (Southern Illinois University Press), and she is co-editor, with Barbara Mirel, of *Reshaping Technical Communication: New Directions and Challenges for the 21st Century* (Lawrence Erlbaum Associates, Inc.). She is also the coordinator of the Research Committee for the Association of Teachers of Technical Writing. Rachel has taught technical communication for two decades, has completed three qualitative studies of workplace writing, and has worked a combined seven years in industry.

The Impact of Student Learning Outcomes Assessment on Technical and Professional Communication Programs

Jo Allen

North Carolina State University

Because of accreditation, budget, and accountability pressures at the institutional and program levels, technical and professional communication faculty are more than ever involved in assessment-based activities. Using assessment to identify a program's strengths and weaknesses allows faculty to work toward continuous improvement based on their articulation of learning and behavioral goals and outcomes for their graduates. This article describes the processes of program assessment based on pedagogical goals, pointing out options and opportunities that will lead to a meaningful and manageable experience for technical communication faculty, and concludes with a view of how the larger academic body of technical communication programs can benefit from such work. As ATTW members take a careful look at the state of the profession from the academic perspective, we can use assessment to further direct our programs to meet professional expectations and, far more importantly, to help us meet the needs of the well-educated technical communicator.

If there is a single hot topic in higher education these days, one that readily embraces so many other competing hot topics, it may well be assessment. In many ways, the assessment movement in American higher education has moved beyond its infancy and at least toward adolescence, if not maturity, in our educational environment (Ewell). The potential impact on our programs in technical and professional communication is significant. Although sometimes resisted, assessment has also proven to be powerfully effective for planning, designing, and promoting distinctive programs and then recruiting desirable students and faculty, as evidenced by the nationally recognized work at Alverno College, University of Central Florida, Indiana University-Purdue University at Indianapolis, Isothermal Community College, and North Carolina State University, among others. The potential of assessment to help with quality initiatives, as well as institutional image and even donor relations, is equally attractive to those who understand, manage, and support quality assessment initiatives and, more important, quality education. As ATTW and its members take a closer look at the role of academics and effective planning, individual programs can also benefit from systematic self-examination of the current state of the curriculum as a means of planning for the future based on data gathered about the student experience and learning outcomes.

Most institutions now have assessment work underway, whether that work was imposed by the state, accrediting bodies, or their own administration. The best scenario of all, of course, occurs when assessment arises from the genuine curiosity of the faculty as a simple question: "Does what we do matter?" In fact, the critical power of assessment depends directly on the *faculty's* taking charge of assessment's processes and opportunities. As an exclusively designed top-down mandate, assessment rarely has a chance to succeed beyond the immediate applications imposed by accreditation (or other mandates). As challenging as assessment may be—given faculty loads and other demands—it is far preferable, in most cases, to the experiences of other countries (and even a few states in the United States) that have mandated outcomes for educational programs. Those results are all too frequently hard to use in beneficial ways for faculty and program planners. Not only do such mandates address imperatives that the faculty may not value, but they also tend to generate useless information that cannot be tied to program improvement. What, for instance, does the number of tenured faculty or volumes in the library actually tell us about a program's quality—especially in terms of whether students know and can exercise discipline-based wisdom and expertise at the conclusion of their studies? In other words, is a program with 100 percent tenured faculty better than a program with 75 percent tenured faculty? How would we determine the answer? With what level of confidence?

Standardized assessment may also fall short because it tends to miss the best features of a program. Its very reliance on common standards for reporting means it caters to the average, as well as promotes a tendency toward checklists ("got that," "do that," "have that") and cookie-cutter approaches for reporting. And we know from nonpositivist arguments about research that insistence on a single set of standards or lens from which to view any kind of evidence typically limits the findings to evidence that is observable from that perspective. Our colleagues in K–12 education quite often voice the same dissatisfaction with end-of-grade or standards testing; the tendency toward mechanized education and concomitant testing is overwhelming—and well worth resisting.

We can do better. In this article, I describe some of the basics of programmatic assessment based on learning outcomes in technical communication rather than on standardized assessment questions as a valuable tool for faculty and program leaders who are looking for evidence of impact and means for continuous improvement. As a faculty member-turned-administrator, I can verify the complexities of assessment-based program work from both sides, but I can also attest to the power of such processes as a means for creating and promoting highly effective and highly attractive programs for students, faculty, employers, alumni, granting agencies, and even donors. In order to hold that kind of meaning, the focus of program assessment must be on the learning that occurs as a result of the curricular and cocurricular experiences of the students. This approach to assessment, widely applied throughout technical communication, can improve the quality of the field overall, as individual programs test or recognize the means to enhance their effectiveness.

A different form of assessment—a needs assessment—is also critically valuable for those who are in the early stages of planning or proposing a new program. Such preliminary assessments are standard in higher education, as program planners look for evidence of the need for the program based on student interest and ability, employer need and support for the graduates of the program, and the institutional strengths to support such a program. As program planners conduct the needs assessment, however, they should also be looking for ways to embed learning outcomes assessment within that program, rather than attaching it as an afterthought, to help them monitor the ongoing needs for modifications as the program is established.

The work of assessment is important to technical communication for two reasons: (1) As part of the academic enterprise, we will be called on (if not already) to provide evidence of our students' learning and abilities; failure to do so may not only threaten institutional accreditation but, just as importantly, may signal that technical communication lacks legitimacy as a true program with meaningful outcomes, and (2) technical communication courses are perfectly situated to engage discussions of both knowledge and ability, making it a natural showcase for accountability and evidence of impact on knowledge/attitudes and skills/behaviors.

Just as important is our understanding that technical communication courses in most institutions are poised to be the primary (if not the sole) scaffold for assessment of learning in the disciplines. So many of our curricular partners now require at least one technical communication course for their students that it is highly likely that faculty will use the work of those courses as evidence of students' writing abilities and content learning. We in technical communication need to set the stage for how that assessment will work best. Doing so requires that we identify our own standards of performance in a technical communication course, the kinds of experiences we want students to have in our programs, and the outcomes we hope to encourage. Otherwise, general standards are likely to be imposed that do not tell us much about the value that our courses are adding to the students' educational experience.

Our colleagues have produced some interesting examinations of course and individual student evaluations/assessments (see, for instance, Bishop; Byrne; Cook; Covington and Keedy; Dragga; Elliot, Kilduff, and Lynch; and Smith) but have rarely ventured into larger programmatic examinations of assessment. In fact, despite some occasional programmatic work (see, for instance, Allen; Anderson; Coon and Scanlon; Coppola), technical communication as a discipline has not embraced assessment fully or in ways one might expect. While the Council for Programs in Technical and Scientific Communication does offer a program review (see www.cptsc.org), that process is more directly tied to performance indicators than to student learning assessment. The special issue of *Technical Communication Quarterly* (*TCQ*) on assessment, Winter 2003, addresses some of those complexities, but we need significantly more work to help us manage the use of writing as evidence of learning for our own programs, as well as for learning in constituent programs across the curriculum. Further, as the National Research Council reports in their study *Knowing What Students Know*, "Educational assessment does not exist in isolation, but must be aligned with curriculum and instruction if it is to support learning" (3). While we have begun to attend to the curriculum part (i.e., what we teach in class), we have done less to attend to the instruction part (i.e., how we teach in class), other than to offer assignments that have worked and means of evaluating them. We have more opportunities, in other words, to use assessment to help us refine our pedagogies for technical communication.

This article, however, focuses just on the first part of that challenge: designing an assessment plan that uses writing as evidence of learning for our own programs' curricular development and improvement. While it is tempting to use individual student evaluations exclusively to measure program effectiveness, such strategies are discouraged because multiple uses of a single measure tend to compromise the strength and veracity of the measure. As high-quality research reminds us, triangulation (the use of more than one method to approach a topic or situation and ferret out information) is substantially preferable in terms of the resulting quality of information and the confidence we place in that information. I hope that others may take up the issue of pedagogical method as a focus of assessment.

WHAT IS ASSESSMENT-BASED PROGRAM REVIEW?

First of all, *assessment* in the context of higher education, and especially in academics, is a process of asking questions about what students get (general knowledge or information, thinking or performance-based abilities, theoretical and applied understandings, practical experience, and so on) out of what faculty and curricular planners provide in the educational arena. Among other possibilities in Academic Affairs, assessment may mean asking questions about (1) students' satisfaction with their educational experience, (2) the amount of their engagement or participation in that experience, or—most powerfully—(3) what they actually learned from that experience. Thus, the most sophisticated forms of assessment focus on student learning outcomes: what students learn and what they are capable of doing as a result of their educational program. Of course, those outcomes may also include what they learn and can do as a result of other educational experiences as well—internships, cooperative education opportunities, undergraduate research

experiences, travel (study or work) abroad, service learning, participation in organizations, and so on. As many will argue, everything counts in constructing the total picture of educational impact and intellectual growth.

These three categories of academic outcomes are easily clarified. For questions
resulting from satisfaction-based assessment, for instance, the aim is to find out
simply what students did and did not like or, more usefully, which aspects of a program, curriculum, or course they did and did not feel benefited them as learners.
For a technical communication program (or even an entire institutional experience), students might be asked to rate their satisfaction with the quality of instruction, the variety of coursework, the level of challenge they felt from their professors and/or peers, the amount or quality of support from tutors, or perceived
support for professional development and job searches.

For engagement or participation-based assessment, students would most likely
check off the sorts of activities related to the program or the entire educational experience in which they took part: athletics, Greek life, professional societies,
travel, internships, work study, research, leadership roles, and so on. For technical
communication students, we might ask about participation in the local Society for
Technical Communication (STC) chapter, tutoring sessions, internships or cooperative education experiences, research, and other forms of deeper educational engagement. That engagement might be even more academically specific, focusing
on class attendance, participation in discussion or collaborative venues, and famil-
iarity with current issues represented in professional journals, for instance. Following that checklist might be questions about the number of hours dedicated to
that activity. Or, students might be asked to rate the degree to which they believe
that such activity made an impact on their learning. Students may report, for instance, that the time they spent ascertaining the finer points of audience analysis
helped them to sharpen a range of related communication skills involving choices
such as vocabulary, tone, structure, exemplification, and graphics that eventually
led them to understand the power of the decisions they made about multiple elements of any given document.

The third category of assessment provides information about student learning
outcomes and is often most significant for faculty and program leaders because it
most closely aligns with our responsibilities to educate our students. This student
learning outcomes approach to assessment asks, "As a result of this program, what
do you now know and what are you able to do?" Assessment experts frequently re-
fer to measures of academic impact as "value-added"—what can the student do as
a direct result of the program that she was unable to do before the program? While
there are some standard measures, and, indeed, the whole notion of pretests and
posttests is predicated on the impact of value-added evidence, experts also point
out that it may be that outcomes themselves are the most significant evidence—
that perhaps we should not care where students get this learning and ability, provided they have it at the completion of our programs. For some, the impact of the

value-added distinction is an important one; for others, it is not so important. Faculty and administrators will need to decide whether they want to capture the location of the value (classroom or other site) as they design their own assessment tools.

QUALITIES OF A GOOD ASSESSMENT STRATEGY

Like most other educational endeavors, a good assessment strategy begins with a purpose and direction:

- What do you want to accomplish with this assessment? Course improvement? Program improvement? Teaching improvement? Student learning improvement? Student employability? Evidence of quality or employability that will attract more students? Better students? More employers? More donors?
- How will information be used, or in what ways do you want this assessment to have impact? On the curriculum? On the program's reputation? On faculty hiring? On student recruiting? On student placement? On the budget?

Again, the tendency is to force assessment to do too many things at once. A more streamlined approach that matches our primary objectives for conducting assessment is best—and the decision about primary objectives should remain in the hands of the faculty with guidance and support of administration.

Based on the works of numerous assessment experts (Astin; Ewell; Polomba and Banta; Banta and Associates; Lopez; and others), a consensus about the characteristics of effective assessment that can lead to an enhanced, high-quality education and even transformation in terms of continuous improvement and data-driven decisions begins to emerge. Such an assessment does the following:

- Asks important questions as agreed on by the program designers and implementers
- Reflects institutional and program mission, where applicable (for example, the compatibility of a particular program with, say, a liberal arts institution)
- Reflects programmatic goals/objectives and outcomes for learning and development
- Contains a meaningful approach to assessment planning as agreed on by those delivering the program
- Is linked to decision making about the curriculum and/or program
- Is linked to processes such as planning and budgeting
- Encourages involvement from individuals on and off campus
- Employs relevant assessment techniques and triangulation of approaches
- Includes direct evidence of learning and/or development

- Reflects what is known about how students learn and/or develop and how faculty/staff learn and/or develop
- Shares appropriate information with multiple audiences
- Shows evidence that data were shared internally
- Leads to reflection and action by faculty, staff, students, and external constituents
- Allows for continuity, flexibility, and improvement in the assessment process
- Provides rewards for those engaged in assessment
- Promotes collaboration across organizational lines
- Reflects a balance of commitment from faculty and administration in providing resources and education to learn and implement the process
- Presents evidence of depth and pervasiveness and sustainability

Clearly, this set of qualities is ambitious but certainly worth striving toward in order to help assessment achieve its potential to affirm a high-quality program or refine its rough edges to improve the curriculum.

EMBARKING ON THE JOURNEY: CLARIFYING THE PROGRAMMATIC CONTEXT

A sophisticated assessment of student learning outcomes can take a number of forms, but it typically begins with asking questions about the program's objectives or goals, questions best directed to the faculty. The faculty will most likely drive this conversation down to the course level at some point, but it is often helpful to begin at the larger, programmatic level. To begin, we may ask a basic question: "What should graduates of our program know and be able to do?" This is no small question, as most faculty who engage in full conversation to answer it quickly learn. Many faculty report the fascinating (and sometimes even surprising) perspectives their colleagues hold on what it means to be a well-educated technical communicator. But the conversation is, nevertheless, a critical first step in the development of a viable assessment strategy. The first step is asking the following series of questions, well-honed from experience and the literature. (These questions appear on the North Carolina State University Web site on academic program review: www.ncsu.edu/provost/academic_programs/uapr/process/overview.html.)

- What are we trying to do, and why are we doing it?
- What do we expect the student to know or to do as a result of our program?
- How well are we doing it?
- How do we know?
- How do we use the information to improve?
- Do the improvements we make work? How do we know?

Deciding on Program Objectives

As early as possible, faculty and program leaders should carefully consider what a graduate from the program should know and be able to do. As they consider these questions, the following opportunities arise:

- To identify specific writing abilities and mastery of particular genres and concerns that will serve as primary evidence of student learning
- To explore more precisely the faculty's view of the roles of technical communicators in society (business, industrial, political, economical, civic) and the curricular artifacts that raise, inform, or enhance those roles
- To enhance relationships with employers who hire the program's students by bringing them into the investigation of the strengths and weaknesses of the curriculum in light of their needs
- To model ownership of assessment for other faculties on the campus
- To model the use of data and analysis as points for defensible decision making related to the curriculum, pedagogy, course sequencing, staffing, recruiting, and other matters that directly impact the quality of the program
- To set the standard for using writing as evidence of thinking and learning

As some examples of what may be desirable outcomes, the faculty might consider the curricular interventions that will best hone their students' abilities to do the following:

- Establish a clear and rich context for their communications, focusing on business, scientific, technological, and/or social drivers for that work
- Articulate the contextual, linguistic, experiential, and/or intellectual background and resulting needs of any given audience for any given piece of information
- Demonstrate the ability to craft technical and professional documentation that builds on particular discourse genres (i.e., instructions, persuasion, description)
- Demonstrate effective editorial and revising strategies that help the writer rework a document, in light of some of these concerns
- Create an effective design for the document that is both appropriate for the content as well as effective for and pleasing to the reader
- Select appropriate media through which to deliver their work to their audience

Some other considerations might include fluency among types of documents, with a variety of audiences, documentation styles, languages, designs, technological tools, management styles, quality measures, production schedules, ethical implications of documentation choices, and so on. But these few, broad strokes might begin to move a faculty through a discussion articulating the qualities and abilities of a well-educated technical communicator.

Moving from Objectives to Outcomes

After making some of these decisions about the bigger picture, the faculty should seek more detailed demonstrations of the development of particular strengths. For instance, in the consideration of students' learning about audience, the faculty might gauge the following kinds of outcomes as meaningful evidence of that learning:

- The ability to articulate differences in audiences' educational backgrounds and their likely effects on comprehension, need for specific types and displays of information, capacity for understanding supporting and contrary evidence, and so on
- The ability to articulate special needs that a general audience has when reading about highly developed technical and scientific topics and the communication strategies for addressing those needs
- Evidence of skill using multiple linguistic, stylistic, and organizational strategies to address a particular audience
- The ability to present a single topic to at least three different audiences with different backgrounds and needs
- The ability to defend communication decisions about content, language, and organization related in the context of a particular audience and genre

You can see that the kind of outcomes and intent are quite different in these examples; the choices the faculty make about such characteristics will, most likely, distinguish the program or some component of it as highly theoretical, highly practical, or some identifiable combination of the two in its goals for student learning.

Collecting Evidence

At this point, the faculty may be ready to collect evidence and adopt appropriate evaluation methods (what they will use as "data" and how they will determine whether such abilities actually exist among their students) and even for points in the curriculum where such opportunities for learning actually exist. With well-constructed assessments that measure student learning and abilities, the faculty have hard data, rather than anecdotal perceptions, of their programs' strengths and weaknesses.

So where does this evidence come from? The easiest answer—and downfall of immature or ill-informed assessment efforts—is that it comes from the grades we already give our students, leaving the impression that there is nothing else to do. In some cases, that impression may be correct. In most cases, however, it is not. The difference is whether the collective faculty feels fully confident in the absolute consistency of their expectations and, thus, consistency of grading across the technical communication curriculum. If just one person believes that faculty emphasize different things in their assignments and in their grading, then reliance on grades as a measure of program quality is unsatisfactory. In addition, if the faculty waver in their conviction that the curriculum has truly covered everything desirable for a graduate from the program, then they may value the perspective that a full assessment would provide. If they think something may be missing, especially given that few programs have lock-step requirements that take every student through exactly the same curriculum with exactly the same faculty member and at exactly the same pace and order, then program review is clearly a valuable addition to traditional course-based evaluations of students' abilities.

The challenge is constructing contexts from which reliable evidence can be collected about the program's strengths and weaknesses. As in many well-designed research projects, triangulation (using more than one approach to provide evidence) is valuable, as are both qualitative and quantitative methods. Some possibilities for gathering evidence of a program's effectiveness include surveys of the students, faculty in the program, and employers about students' abilities, knowledge, confidence, attitude, and initiative, for instance. To generate more specific evidence of learning outcomes, internships or cooperative education experiences that provide work-related assignments and documents are a fine source of evaluation. The additional benefit of gathering employers' feedback on that work is the legitimacy of the workplace perspective that employers provide, as well as the legitimacy of the students' writings as "real-world" assignments. And, of course, capstone courses provide an opportunity for students to demonstrate what they have learned throughout the program, rather than in a single course. Students' self-assessments may also be valuable, provided they are accustomed to the intent and use of such assessments. In all of these situations, carefully guided observations and commentary, based on clear criteria, are critical for a meaningful assessment.

USING STUDENT WRITING AS EVIDENCE
OF LEARNING

In considering options for assessment, we begin by looking at readily collectable evidence. What do we already have at hand that we can use to determine the existence and quality of learning outcomes? Many faculty look at direct evidence of student work—their writing—most often collected in a portfolio as the corpus of evidence from which to gauge effective instruction and learning. Key, however, is that we do more than simply collect portfolios or writing samples. To be useful, we must confront the evidence, testing it to determine whether it substantiates our claims about teaching and learning and, ultimately, whether our program has met its goals. Consequently, we must (1) identify points in the students' education where those criteria should have been introduced and honed and (2) articulate the criteria against which we will analyze that evidence or data (the writing samples) that our students provide.

Matrices: Points of Intervention and Learning in the Curriculum

Some faculty have, at this point, created matrices that allow them to chart points in the curriculum where specific topics and abilities are introduced or developed (See Table 1 for an example.) Such a matrix allows the faculty to see and to agree on the points of development in their students' academic program where specific abilities should be more evident. It also allows the faculty to engage in formative or intermediary evaluations of their entire student population to see how the program is doing at specific milestones, such as the end of the junior year, or midway through the senior year. Faculty can ask, in other words, whether the program encourages the kind and amount of growth in ability and understanding that it anticipates at

TABLE 1
Curricular Matrix

	ENG 301	ENGL 302	ENGL 305	ENGL 311	ENGL 401	ENGL 441
Genre development						
Instruction		x				x
Summary	x					
Data analysis				x	x	x
Problem-solution	x	x	x		x	x
Literature review	x					x
Proposal			x		x	x
Memos/letters	x			x		
Comments:						
Skills development						
Summary	x		x			x
Analysis	x	x				
Problem-solving	x	x	x	x	x	x
Critical thinking	x	x	x	x	x	x
Persuasive writing		x		x		
Synthesis	x					
Audience analysis	x	x	x	x	x	x
Design analysis		x		x		
Software manipulation				x		
Text edits	x	x		x		
Qualitative research			x		x	
Quantitative research			x		x	x
Comments:						

those pivotal intervals. If so, great. They now know that their program is meeting its goals and graduating well-educated students, and equally, if not more, important, they can articulate what a well-educated student in their program actually knows and can do—and they have firm evidence of that learning.

If not, of course, they should determine what needs to change as well as where in the students' experience and when. Do changes need to be made at the course-level or at the program level? At the sophomore, junior, or senior level? In short, do faculty need to revise a course, or do they need to add more courses or infuse more of the curriculum—rather than a single course—with opportunities for students to learn and practice those desirable skills? Would a practicum, internship, or cooperative education experience help? Would a new teaching strategy help? A new or clearer assignment? Would collaborative work be more advantageous?

Rubrics: Criteria and Analysis for Evaluation of Evidence

While the matrix provides a checklist of points in the curriculum where certain teaching occurs, that is not to be confused with an assessment of whether students are actually learning at those points. For that, we need the hard evidence—the students' writings that serve as our data in this kind of assessment. And as we look through the students' writings (again, most often collected in a portfolio, intermediary, or capstone experience), we must find a meaningful way to establish criteria and the degree to which the writings meet the criteria. Most useful at this point is a *rubric*, a simple guide that allows evaluators to chart a particularly desirable characteristic or set of characteristics against a continuum of the degree of its existence in the data. It's far simpler than it sounds. Nancy Coppola provides a sample of a rubric for evaluating a portfolio, looking for specific evidence of global characteristics of good writing (254–56). Table 2 provides a sample of a rubric for evaluating an individual or collective sample of a specific genre: persuasive writing. Readers (faculty, external reviewers, employers, advisory board members, or others) may read works and measure the level of competence or mastery with each characteristic, helping to draw a clear picture of where students' writings are strong and where they could improve. Those judgments feed directly back into the matrix, created earlier, that identifies particular points in the curriculum (or in a course) where certain skills should have been learned. Once the program's faculty or administrators see that, for instance, students have difficulty closing a persuasive piece or moving between points in making their case, then the faculty have a better idea of what to do the next time they offer the course(s). This evidence and response is critical to continuous improvement and demonstrates the practice of "closing the loop," an assessment metaphor for using evidence to make decisions about course and program improvements and then evaluating the impact of the change.

TABLE 2
Persuasive Writing Rubric

Category	Exemplary	Developing	Beginning	Level Assigned
Knowledge of techniques or strategies of persuasion	Clearly demonstrates knowledge of numerous techniques and options for creating persuasive text (i.e., direct challenges, metaphor/comparison, sarcasm, humor, etc.)	Demonstrates some knowledge of a couple of techniques	Demonstrates no knowledge of variety of techniques	
Application of techniques	Clearly demonstrates mastery of applying various techniques to promote the persuasive mission of the text (i.e., direct challenges, metaphor/comparison, sarcasm, humor, etc.)	Demonstrates mastery of one technique but no fluency with others	Demonstrates little to no familiarity with the means to apply any/all techniques	
Audience	Clearly demonstrates appropriate consideration and accommodation for audience's background and stance on the topic	Demonstrates some consideration for audience's background and/or stance on the topic but is not thoroughly integrating those concerns with the text	Demonstrates little to no familiarity or awareness of the audience's background or stance on the topic	
Tone	Clearly demonstrates mastery of the tone of the persuasion, as appropriate for the audience and topic	Demonstrates some ability to manage the tone, as appropriate, for the audience and topic	Demonstrates little to no ability to manage the tone of the work, as appropriate for the audience and topic	
Organization	Clearly demonstrates an understanding of the best organizational approach for presenting the persuasive argument for the audience and topic	Demonstrates some understanding of organizational preferences but makes some errors in judgment/implementation	Exhibits no ability to organize appropriately for the audience and topic	

(continued)

106

TABLE 2 (Continued)

Category	Exemplary	Developing	Beginning	Level Assigned
Analysis of the topic	Clearly explains the challenging issues raised by the topic and acknowledges research, statistics, and findings related to the topic	Somewhat explains the challenging issues raised by the topic and acknowledges some of the research, statistics, and findings related to the topic	No explanation is provided of the challenges raised or dissenting views	
Synthesis of points	Exhibits the ability to integrate the findings related to the topic from multiple sources	Exhibits some ability to integrate the findings related to the topic from two or more sources	Exhibits no ability to integrate the findings of the topic; relies entirely on serial chronicles of findings	
Evidence of persuasive ability	Able to articulate well-reasoned and well-crafted objections to contrary opinions or evidence	Can challenge some contrary opinions and evidence but with little confidence or level of mastery	Can challenge no contrary opinions and evidence; or excludes contrary opinions and research from discussion	
Conclusion	Clearly moves the audience to the reasoned conclusion stated in the document	States a conclusion but doesn't take audience through the reasoning to that conclusion	Ends the work inappropriately, without making the final point or providing the point of conviction	
Comments:				

EVIDENCE OF A MEANINGFUL LEARNING ENVIRONMENT: BEYOND LEARNING AND OUTCOMES

While this discussion of program review highlights the opportunities and directions the faculty might take with such an assessment-based approach, we must remember that the curriculum is only one element of a meaningful learning environment and experience. At some point, faculty may want to investigate their institution's or program's contributions to the other elements that are known to impact student learning, engagement, and, ultimately, satisfaction and persistence: advising, faculty mentoring, opportunities for undergraduate research, opportunities for leadership, participation in social events, participation in professional organizations and athletics. In addition, they may want to look at intersections of financial aid, scholarships, course-taking patterns, academic support, and learning styles as they also influence learning and satisfaction.

Finally, from the larger perspective, we might also consider how our profession would benefit if all programs clearly articulated their goals for their students' educational outcomes (knowledge + abilities) and specified evidence of those objectives/goals and outcomes. And we might consider how advantageous it would be to publicize that information in the Web sites and marketing brochures we so beautifully churn out. For one thing, we could advance the legitimacy of our discipline, especially in response to contentions that our work is "subjective." For another, we could begin to distinguish between various programs and their primary emphases and philosophies. We could, with greater authority, advise students about their graduate options, based on the emphases of programs throughout the nation—whether theoretical/applied, technology/science/business, or other distinctions. With that confidence, we could also use our programs' differences as evidence of the richness of our discipline, rather than as suggestions of fragmentation or even confusion. In short, the clarification of what we want a graduate of our program to know and do would mark both the individuality and the common ground that advances twenty-first-century technical communication as a true discipline.

WORKS CITED

Allen, Jo. "The Role(s) of Assessment in Technical Communication: A Review of the Literature." *Technical Communication Quarterly* 2 (1993): 365–88.

Anderson, Paul V. "Evaluating Academic Technical Communication Programs: New Stakeholders, Diverse Goals." *Technical Communication* 42 (1995): 628–33.

Astin, Alexander W. *Assessment for Excellence: The Philosophy and Practice of Assessment and Evaluation in Higher Education.* New York: American Council on Education, 1991.

Banta, Trudy, and Associates. *Building a Scholarship of Assessment.* San Francisco: Jossey-Bass, 2002.

Bishop, Wendy. "Revising the Technical Writing Class: Peer Critiques, Self-Evaluation, and Portfolio Grading." *The Technical Writing Teacher* 16.1 (1989), 13–25.

Byrne, Ros. "Written Feedback on Student Assignments: Another Look." *Business Communication Quarterly* 60.2 (1997): 100–108.

Cook, Kelli Cargile. "How Much Is Enough? The Assessment of Students in Technical Communication." *Technical Communication Quarterly* 12 (2003): 47–65.

Coon, Anne C., and Patrick M. Scanlon. "Does the Curriculum Fit the Career? Some Conclusions from a Survey of Graduates of a Degree Program in Professional and Technical Communication." *Journal of Technical Writing and Communication* 27 (1997): 391–99.

Coppola, Nancy W. "Setting the Discourse Community: Tasks and Assessment for the New Technical Communication Service Course." *Technical Communication Quarterly* 8 (1999): 249–67.

Covington, David H., and Hugh F. Keedy. "A Technical Communication Course Using Peer Evaluation of Reports." *Engineering Education* 69 (1979): 417–19.

Dragga, Sam. "Responding to Technical Writing." *The Technical Writing Teacher* 18 (1991): 202–21.

Elliot, Norbert, Margaret Kilduff, and Robert Lynch. "The Assessment of Technical Writing: A Case Study." *Journal of Technical Writing and Communication* 24 (1994): 19–36.

Ewell, Peter T. "An Emerging Scholarship: A Brief History of Assessment." In Trudy W. Banta and Associates (Ed). *Building a Scholarship of Assessment*. San Francisco: Jossey-Bass, 2002. 3–25.

Lopez, Cecilia. *Opportunities for Improvement: Advice from Consultant-Evaluators on Programs to Assess Student Learning*. North Central Accreditation Commission on Institutions of Higher Education, 1996.

National Research Council. *Knowing What Students Know: The Science and Design of Educational Assessment*. Washington, DC: National Academy Press, 2001.

Palomba, Catherine A., and Trudy W. Banta. *Assessment Essentials: Planning Implementing, and Improving Assessment in Higher Education*. San Francisco: Jossey-Bass, 1999.

Smith, Summer. "What Is 'Good' Technical Communication? A Comparison of the Standards of Writing and Engineering Teachers." *Technical Communication Quarterly* 12 (2003): 7–24.

Jo Allen, President of ATTW 2003–2005, is Interim Vice Provost for Undergraduate Affairs at North Carolina State University, where she oversees ten programs and offices (including undergraduate program assessment), as well as the planning and budgeting that promote special academic initiatives for undergraduate education. She has published books and papers and presented on higher education and communication issues in over one hundred scholarly venues, has served on four editorial boards, and has consulted extensively regarding technical and administrative communication.

Reflections on *Technical Communication Quarterly,* 1991–2003: The Manuscript Review Process

Mary M. Lay
University of Minnesota

This article traces the development of *Technical Communication Quarterly* (*TCQ*), beginning with the first issue in the winter of 1991, through the 2003 issues. As co-editor of *TCQ*, charged with the manuscript review process, I shepherded more than 350 manuscripts through evaluation and about one-fourth of those through publication. In this article, I explain that process and how it changed when *The Technical Writing Teacher* became *TCQ* and what features our reviewers now believe make a successful *TCQ* article.

I cannot remember whether it was at a meeting of the Council for Programs in Technical and Scientific Communication or of the Conference on College Composition and Communication (CCCC) when Billie Wahlstrom and I first discussed the status of the official journal of ATTW, *The Technical Writing Teacher,* and what changes we would initiate as co-editors of *Technical Communication Quarterly* (*TCQ*). We were on the beach (so it must have been either Orlando or San Diego), and we were excited and honored to have been chosen by the ATTW Executive Committee to edit the journal. I do remember the long discussions that ATTW had about the journal over the four-year period in the late 1980s and early 1990s when first I and then Jack Selzer led the association as its presidents. ATTW members were calling for several changes: The name of the journal no longer represented who we were and what we did; the cover and binding needed updating; and we wanted to see more interdisciplinary and theory-based articles. ATTW members concluded that we were not only teachers but scholars and researchers as well. And our research and teaching focused not only on writing but also on speaking and on visual design. Jack suggested that we publish four issues a year, and so we became the *Technical Communication Quarterly* (we all liked the sound of *TCQ*). Building on the fine work of the past editors, Don Cunningham and Victoria Mikelonis, we wanted *TCQ* to be a journal that scholars and teachers would seek first, and that tenure and promotion committees would find impressive. Finally, Iowa State had set a high standard with the *Journal of Busi-*

ness and Technical Communication, and we needed to compete with them but distinguish ourselves from them. So Billie agreed to handle production, and I agreed to handle manuscript review—and *TCQ* began.

In this article, I focus on the manuscript review process. I explain the process itself and then spend the bulk of this article on what reviewers seek in a manuscript—in particular, what causes the two extremes in evaluation: rejection or enthusiastic acceptance. I read through the reviewers' comments for the last two years (2001–2003) and pulled representative comments, masking the identity of the reviewer and author and any markers of the content of the review that might too closely reveal which manuscript elicited which reviewer comment. *TCQ* reviewers are very generous in volunteering their time and energy, and here again they were gracious in allowing me to quote or paraphrase from their reviews.

Manuscript reviewers for *TCQ* either nominate themselves or are suggested by current reviewers; about half are established scholars and half are relatively new to the profession. They agree to serve a three-year period and usually renew for a second or third term. They identify which content areas or topics they consider their specialty and are assigned manuscripts in that area. These topics, to a great extent, mark what we believe represents reader interest and are updated every two to three years. The topics at the time of this writing are listed in Table 1.

The majority of manuscripts received in the past two years fall primarily into the topics marked with a dagger (†) in Table 1. Although few, if any, topics have been dropped during the past ten years, several have been added; those added most recently are marked with an asterisk (*). For example, when the first special issue on gender in technical communication was proposed, it was clear that we needed reviewers who considered this topic a specialty. Certainly the advent of distance learning and online course work produced articles that needed specialists to review them. Reviewers themselves added topics in the "other" category on their reviewer agreement forms. On the other hand, although some topics remain on the list, we seldom receive a manuscript that would fall into those categories. For example, business communication articles are probably now submitted elsewhere. Pedagogy issues as a category has been refined to specific topics, such as distance learning, or graphics and visual communication. Curriculum assessment may be part of the conclusion of an article that suggests a specific pedagogical direction or presents the results of classroom research. Even though many technical communication instructors teach engineering writing, and technical communication scholars study workplace writing, we haven't received a manuscript that addressed this topic exclusively in several years. However, the list does provide a reflection of what kinds of topics are addressed in the journal, and, in choosing the appropriate reviewers for a manuscript, the list compared to the abstract and works cited section of the manuscript helps us make a good match. Of course, the editor must do some interpretation. For example, because Catherine Fox applies Burke's theories to a case study of technical writers in the workplace, should her article "Beyond the 'Tyranny of the Real': Revisiting

TABLE 1
Topics of Articles in *Technical Communication Quarterly*

Advertising/marketing	Hypertext/hypermedia*
Audience analysis†	International/intercultural communication*†
Business communication	Language/learning theories
Collaborative writing†	Law and/or liability*
Communication theory†	Medical writing/communication*
Computers/computer technology†	Multimedia education and documentation†
Designing/developing programs in technical writing/communication†	Nature of technical communication research†
Discourse analysis†	Organizational theory/communication
Distance learning*†	Pedagogical issues†
Documentation in corporate cultures†	Program/curriculum assessment and/or evaluation†
Editing	Publications management†
Engineering writing	Qualitative studies†
English as a second language	Quantitative analysis†
Environmental issues*†	Rhetoric of science and technology*†
Ethics in technical communication*†	Rhetorical theory/invention†
Gender or feminist studies*†	Systems theory
Graphics/visual communication†	Telecommunication†
History of rhetoric†	Usability testing*†

Note. †indicates one of the topics which accounts for the majority of manuscipts received during the past 2 years. *indicates a topic that has recently been added for consideration.

Burke's Pentad as Research Method for Professional Communication," published in the Fall 2002 *TCQ*, be read by specialists in rhetorical theory, nature of technical communication research, documentation in corporate cultures, or all three? Reviewers have the opportunity to turn back a manuscript if they think it falls outside their area of expertise, but sometimes a reviewer will be asked to read as a representative of the general reader, while the other two reviewers read as experts in the topic. A further revision of the list (to be done at the end of 2005, when the new editors renew contracts with the reviewers) should include such topics as consulting and service learning, two topics addressed in the Fall 2002 issue of *TCQ*.

When several scholars are clearly addressing a new topic, one of those scholars may propose that a special issue of *TCQ* publish four to six articles that together reflect the new research. In essence, the special issue tries to capture in the articles and in the special editor's introduction the origins and nature of the conversation about that topic. The special issue then can provide an in-depth look that a single article published in isolation cannot. The special issue editor selects his or her own set of reviewers and solicits abstracts or manuscripts for those reviewers to evaluate. Those reviewers, along with the special issue editor, work with authors to develop each article. The rate of acceptance is often higher for special issues, which encourages prospective authors, and readers enjoy finding a collection of articles on a particular topic located in one place. Moreover, special issues of *TCQ* have led to anthologies published by scholarly presses. Special issues became important to *TCQ* about ten years ago, and now readers can expect to see one or two a year.

When reviewers receive a manuscript, they are asked to consider several questions in completing a review (see Table 2). Three reviewers read each manuscript, and each reviewer receives a copy of the other reviewers' comments, a process that over the years has formed a community dialogue among reviewers. This community was created early in *TCQ*'s history when reviewers met during a meeting of the CCCC to offer suggestions to the editors and to hear our standards for a thorough review. The reviewers recommend eventual acceptance for between 20 and 30 percent of all manuscripts submitted. About one-third of these articles accepted (that 20 to 30 percent) are accepted conditionally, based on suggested revisions that the editor alone oversees. The other two-thirds of those accepted are revised for reconsideration and are read and evaluated by the original reviewers a second time. Although an article that is revised for reconsideration and evaluated a second time enjoys no guarantee of eventual acceptance, only one or two a year that fall into the "revise for reconsideration" category are advised to submit elsewhere after that second reading. Thus, the author might face a major revision two or three times before seeing his or her article in print, but revision is well worth the effort.

Regardless of the formal guidelines given to *TCQ* reviewers, however, their comments have more impact on the nature and quality of the manuscripts published and the ways in which the manuscripts are revised. Probably the most important feature that reviewers identify in a manuscript recommended for publication—or the first suggestion for revision—is *novelty*. Reviewers most frequently comment on novelty when answering these review questions: Is the topic relevant and of interest to *TCQ*'s

TABLE 2
Questions for Manuscript Reviewers

Is the topic relevant and of interest to *TCQ*'s audience?
Is the article logically organized?
Does it have a clearly stated purpose and adequate introduction?
Does the body of the text support the premise, and is there a balance of theory and research to back
 up the claims?
Is the conclusion accurate?
Does the title reflect the content?
Has the text been adequately researched?
Are the claims backed up by pertinent research and reference material?
Are there any gaps in the text?
Are assumptions made that need explanations?
Are there inconsistencies and grammar and spelling mistakes?
If the article has illustrations, are they clear and relevant to the text?
Are the illustrations suitably presented and of a sufficiently high quality to reproduce?
Are the references adequately documented (and in MLA style)?

audience? Has the text been adequately researched? And are the claims backed up by pertinent research and reference material? The reviewers check to see if the manuscript presents a new look at a topic within our scholarly conversations and if the manuscript emerges from this continuing scholarly conversation.

Carol Berkenkotter and Tom Huckin found the same requirement when they traced reviewer comments on a typical scientific article, and they defined novelty as follows:

> The concept of *novelty*, as it relates to scientific discovery, refers to the idea that innovations (new postulations) are at the heart of the scientific enterprise as it is seen by its practitioners. If scientific activity is to be purposeful and cumulative, then a major criterion for publication is the novelty and news value of the researchers' knowledge claims, seen in the context of accumulated knowledge. (47, emphasis in original)

The scientific writer that Berkenkotter and Huckin shadowed was asked by her reviewers to "foreground her claims" to establish novelty (59).

TCQ reviewers quite frequently make comments on manuscripts such as these: "I don't see much that is new in the solutions that are offered. All of them seem to be techniques that instructors of technical communication are (and have been) using." Or,

> I came away from this piece feeling that the approach the author outlines is not quite as novel as he/she purports. Also, there are existing works that actually have done much of what the author seems to claim is brand new. I was somewhat troubled, for example, that there were no citations of [noted scholar's] work.

Reviewers might ask authors to create novelty in two ways: do research that tests, extends, or creates knowledge, or take a position that might be counter to claimed knowledge. The first position is represented in the following comment:

> This approach seems to have a great deal of potential, especially for helping researchers in our field consider their findings from multiple angles and perspectives. I also like how the author demonstrates this approach by using his/her own research. The reader comes away from this piece with a clear sense of the approach and what it can contribute and with some very concrete ideas about how to use it.

The reviewer comments that follow reflect the second position:

> While the issue of gender and technical writing is an interesting one, this article doesn't add anything to what has already been said in the literature. It would have far more value and interest if you were able to add new research or insight, or perhaps take a "devil's advocate" position.

Moreover, *TCQ* reviewers expect that any claim to novelty would be established by providing what Berkenkotter and Huckin identify as an "intertextual framework for local knowledge" (59). Authors need to know the professional conversation about their topic, not only to establish ethos to talk about that topic, but also to claim that their specific research and thinking extends or refines that conversation. Thus, reviewers might say something like the following:

> This study is well situated within the quite extensive conversation that has gone on in the field over at least the past ten years about the need for more research—and research that is relevant to practitioner needs and experiences. This conversation is fairly central to the ongoing conflict and tension between academics and practitioners and the perpetual question of the relative merits of theory/practice.

Part of capturing that scholarly or professional conversation comes from knowing the audience of the journal—who they are and what they know already about the topic. As one reviewer suggested, "The article seems addressed to English teachers who are new to and untrained in teaching technical writing because of writing across the curriculum assignments. I think that portion of *TCQ*'s audience is very small at best." Thus, one cause for praise by reviewers would be the establishment of novelty within the context of the ongoing discourse about the topic of the manuscript and a recognition of the current knowledge among journal readers, as reflected in the following reviewer comment: "This piece gets off to a good running start, establishing a clear and interesting thesis early, and offers a well developed view of the importance of that thesis to the journal readers."

The reviewers also insist that authors clearly establish the problem they are trying to solve and the methodology and method that they use to solve this problem. The problem must be of importance to *TCQ* readers, as reflected in the following reviewer comment:

> I don't see much evidence that the "problem" identified in the article is having a serious or substantial negative impact on either students or on the readers of their documents. Thus, even if I accept the situation that the article identifies as a problem, I don't have much motivation to try the proposed solutions. How serious is the problem? Why is this problem more important than the other problems that instructors of technical communication are trying to solve?

One of the ways to establish such a problem is to do a thorough literature review, but reviewers usually reject an article if it remains primarily a literature review, as reflected in the following comment:

> I see the article basically as a lit review, and it is well written and researched, but in the last analysis, I do not believe that this literature holds much interest for the reader-

ship. The literature is rife with generalizations about the topic but there is little sub-stantiation in the way of data, empirical studies, and the like. It is not clear how this literature should be evaluated, or how the theory has been or could be tested. The author does not evaluate it for credibility, reliability, nature of arguments, and support-ing data.

Although the reviewers have recommended that a few annotated bibliographies be published in *TCQ*, they have required that the author both describe and evaluate the literature presented and that the literature be new and of major importance to the journal reader. Of course, the literature review presented in the beginning of an article helps establish the author's ethos, and because the reviewers are experts in the subject area, they look closely for any gaps, as in the following comment: "I think that the omission of all of the following studies [a list was provided] is dam-aging to the article's credibility and its conclusions."

Most published articles report on some sort of original research, and, if it is not offered, the reviewers wonder why, as reflected in this reviewer's comment:

> As it stands, however, the paper seems very thin to me. You have identified a general claim but the evidence you provide for this claim is primarily secondary in nature—based on the work of [a list of cited scholars was given by the reviewer]—but you provide no evidence of your own to support your claim. As a reader, I am interested in learning about the scholarly research you have done that led you to believe in the claim that you made.

The reviewers, moreover, look closely at the methodology and method used to complete and interpret the research. For example, they often comment on sample size in a study (a matter of method):

> My main concern is with the single subject case study and how the author uses activ-ity theory to draw some pretty large generalizations. In addition, when it comes to ex-plaining some of the "outcomes" of the research project through activity theory, it al-most seems too neat, too clean. It almost appears that the theory was "sitting there" and that the author found a single subject to fit the theory.

Reviewers also examine if the data support the author's interpretation (a matter of methodology), as reflected in the following comment:

> One aspect of the author's methodology also troubled me. He/she mentions that Y and Z were not available for interviews; however, later in the article a good deal is at-tributed to them. This raises some issues about the reliability and validity of some of the author's claims.

And the reviewers check to see if the author's interpretation over or under reaches that data:

> On the other hand, it appears your data may be richer than your analysis at this point. There are a number of concepts represented in the tables that are not discussed in the article at all. A fuller analysis of your data might be productive.

TCQ reviewers recognize that much scholarship within the field is interdisciplinary, a particular challenge for authors who must adapt theory and research from a discipline such as psychology, history, rhetoric, or anthropology, and present a coherent and valid report for journal readers. The reviewers resist an article that simply says we should pay more attention to this other field and the research going on in it—a consciousness-raising exercise. Instead, they look for a thorough understanding of the outside field and recognition of those technical communication scholars who have already ventured into that field. In other words, they are suspicious of claims of novelty for interdisciplinary work only, as reflected in the following reviewer comment:

> In the study, the author goes outside of traditional technical communication genres to examine the evolution of new media over several decades. I applaud this interdisciplinary work. It might be useful to cite similar technical communication studies that deal with similar comparative historical examinations of technical communication genres.

Likewise, the reviewers may reject an article that claims novelty because a new technology comes along or that is based solely on reporting how that technology was used in the classroom or workplace:

> The article really is not right for *Technical Communication Quarterly*. Technical communication teachers and practitioners are concerned with writing, designing, and producing technical, scientific, and medical communication. This article basically describes the effectiveness of implementing a particular technology.

This is not to say that reviewers disfavor the article that presents practical application. In fact, they prefer that an author ground a theory in practice, as the following reviewer notes:

> Whereas some *TCQ* submissions may underemphasize or fail to address the practical applications of theory in the classroom—this manuscript does address the application and does so well. The humorous examples, graphics, tone all strengthen the manuscript's readability.

The reviewers also examine the overall organization and focus, evidence, and the complexity of each manuscript. For example, one reviewer requests that the author revise for coherence:

> The movement of the piece from reporting a research study to pedagogical recommendations is quite abrupt, and that section seems coarsely written. Smooth that transition and treat as paragraphs in a coherent, smoothly developed section rather than as a welter of points.

Another reviewer finds that the author has too much information for one article, thus detracting from focus:

> I believe that the author could develop publishable studies from at least three of the lines of inquiry sketched in this manuscript. However, I do not believe that the manuscript is unified or fully researched enough to justify publication as it stands.

If an author cannot support a claim, a reviewer might recommend that this section of the manuscript be developed or deleted, as reflected in the following comment:

> Deletion of this section and its sweeping coverage of a variety of tangential issues (each of which could merit explanation and elaboration) would lead the reader more quickly and directly to the heart of the article. I am afraid that readers will be mired in this section, questioning or challenging its unsupported claims instead of proceeding to the later and better sections of the article.

TCQ reviewers, however, want to ensure that in achieving coherence and focus that an author does not oversimplify. The following reviewer resisted an author's attempt to define all the literature reviewed as equally applicable to his or her subject:

> Another problem with this section of the manuscript arises from the author's apparent attempt to argue that the literature on the topic all focuses on the question of whether technical communication should define itself in terms of a "service paradigm." This sort of simplification saps the energy, diversity, and creativity from the actual discourse about which the author is writing.

Finally, if the reviewers give top priority to novelty—the expectation that authors will have done some original research, have presented and solved a problem, or have awakened readers to the importance of a new theory or an interesting application of an existing theory—again they usually frame their comments in terms of fit to the *TCQ* audience. The reviewers are established readers of the journal and define themselves as surrogates for all readers of *TCQ*. They expect that authors of published articles will know the journal well and will demonstrate that knowledge within their manuscripts. They are suspicious of authors who appear to be "shopping around" for placement without doing their research on the journal. One final comment from reviewers illustrates how consistently this requirement comes up. This reviewer's recommendation for acceptance argues for the benefits *TCQ* reviewers will realize from the article:

> The subject is appropriate to the audience of *TCQ*, the authors situate their discussion within relevant scholarship in our field, and they nicely introduce the reader to scholarship from related disciplines. Most importantly, the authors argue reasonably and forcefully for an important change to existing practice. They use good examples of well-informed teaching units to underscore their approaches to instruction, and they are alert to the social and political context of the classroom.

The good news in this assessment of what *TCQ* reviewers look for in a successful manuscript is that they provide two ways to help an author. First, not only do they provide the kind of global comments that I have just documented, but they also offer page-by-page, if not line-by-line, reactions to many manuscripts. These thorough comments frequently ensure that an author who follows them will eventually see his or her manuscript in published form. Second, reviewers are quite willing to read and reread a manuscript, going through a revision to see how the author has incorporated their comments and assessing what final points the author needs to strengthen. They, more than I, have developed those 350 some manuscripts that we saw between 1991 and 2003. An author would be hard pressed not to take advantage of this generosity and expertise.

Over the past thirteen years, *TCQ* has become a leading journal in the field of technical communication scholarship. Manuscript reviewers have set the standard for publishing those articles that have a solid theoretical base and contribute new knowledge. Pedagogical recommendations must be backed up with research in the classroom, be grounded in an understanding of communication or rhetorical theory, and be applicable in classrooms in more than one institution. The primary mission of *The Technical Writing Teacher*, to provide classroom exercises and suggestions, has been turned over to the ATTW *Bulletin* and to the many textbooks now available on technical communication. Although many or most *TCQ* articles include pedagogical implications, all research, including research situated in the classroom or workplace, is developed according to well-identified methodologies and extends, tests, or refutes major theoretical stances. Thus, technical communication scholars have become experts on Burke, Foucault, de Certeau, and others. A good portion of *TCQ* articles offer historical research on the texts produced decades, if not centuries, earlier; extend our research base by drawing on other disciplines; raise ethical issues about our scholarship and teaching; challenge us to recognize the global marketplace in which our students will work; and debate the place of emerging technologies in the workplace and classroom. *TCQ* has positioned itself as perhaps the most theoretically based journal within technical communication, distinct from so-called practitioner's journals such as *Technical Communication*, from those journals embracing both technical and business communication such as the *Journal of Business and Technical Communication*, and from those that share the some theories but not the same research site such as *Rhetorical Society Quarterly*. To a certain extent, *TCQ* helps drive the field by its regular and special issues, starting conversations that will be reflected in future articles, dissertation, anthologies, and books. Its articles appear in instructors' manuals for leading textbooks, because now instructors are expected to know and contribute to this scholarly base if they are to succeed in promotion and tenure reviews. These changes have been risky to some extent. Readers initially, and may continue to, feel that the place for pedagogical debate has shrunk, that some theories are too lofty to be applicable to the classroom and workplace, and that scholars and practitioners do not converse enough. However, *TCQ* has reached its goal of being of such quality that it is not questioned in promotion and tenure reviews and that its reviewers reflect the standard in a scholarly discourse community. Those standards are reflected in the manuscript reviewers' demand that published articles offer novelty on a topic either in the form of new research or a position contrary to what is considered established scholarship and that this novelty be grounded in the existing work in the field.

WORK CITED

Berkenkotter, Carol, and Thomas N. Huckin. *Genre Knowledge in Disciplinary Communication: Cognition/Culture/Power.* Hillsdale, NJ: Lawrence Erlbaum Associates, Inc., 1995.

Mary Lay is a professor in the Department of Rhetoric at the University of Minnesota, where she directed the graduate program from 1997 to 2001 and the Center for Advanced Feminist Studies from 1994 to 1997. She is author of *The Rhetoric of Midwifery: Gender, Knowledge, and Power* (Rutgers, 2000) and co-editor of several anthologies, including *Research Methods in Technical Communication* (Greenwood/Praeger Press/ATTW series, 2002). She has published in *QJS, JBTC, JAC,* and other journals. As well as serving ATTW as manuscript review editor of *TCQ*, she is a past president and fellow of the organization.

The Founding of ATTW
and its Journal

Donald H. Cunningham
Auburn University

Don Cunningham, the founding editor of *The Technical Writing Teacher* and a founding member of ATTW, recalls key moments in the history of ATTW and its journal, and the people who shaped the organization in its early years.

ATTW was launched at a March, 1973 Conference on College Composition and Communication (CCCC) in New Orleans. None of us present thought about it as the first ATTW meeting—we didn't have a name or a title for the journal then, and an organization was not yet even a twinkle in our eyes. Because none of us was aware that we were laying the keel for the organization and all the subsequent developments that followed, nobody was taking notes or attempting to remember exactly what was transpiring. We were simply a small clutch of colleagues noodling over some recurring problems related to establishing quality in teaching technical writing. Although we hoped that our work would have a positive effect on preparing teachers to teach technical writing, we could not have forecast the things that followed. We were not seeking to cut the ground from beneath anybody, although we did want the "furniture" to be rearranged in the National Council of Teachers of English (NCTE) and CCCC conferences to allow for more sessions on technical and business writing. That was about all we were aiming to achieve.

I am not 100 percent certain about some of the dates that I report, although I recall pretty well what occurred and who was there. But I will try to stick to the things that I think are true and cover only those things that I experienced firsthand—counting on a "Hey, I was *there!*" credibility. John Harris, Nell Ann Pickett, Ann Laster, and I may be the last surviving members of the group at that first meeting. John Walter and Herman Estrin were there, but they have since passed away, and I don't think they ever wrote down their memories of that first year.

THE CONTEXT FOR THAT FIRST MEETING,
AND WHAT IMMEDIATELY FOLLOWED

The first time the need for an organization for teachers of technical writing was mentioned in public was at one of the technical writing sessions at the CCCC meeting in New Orleans in 1973. I cannot recall the session topic, although I was a speaker and Herman Estrin chaired the session. I believe that John Harris and John Walter were also speakers.

During the discussion period after the presentations, one of the faculty from the U.S. Air Force (USAF) Academy questioned me hard about some things that I had said in my presentation, especially about teaching such routine documents as application letters and resumes, inquiry and response letters, and instructions. Others in the audience—mostly people who were beginning to teach technical writing or were interested in seeing what it was all about—countered by saying that what I was talking about was exactly what they needed to know. I was feeling roughed up quite a bit before I realized that the questioner was playing devil's advocate and trying to get others in the audience to ask questions. There were the ever-predictable questions from the attendees about how to get help in teaching technical writing, about sources, textbooks, and so on. I used to get calls and mail like that every month during the mid- to late 1960s and all through the 1970s. The field lacked established forums for sharing of information, and people sought help in an ad hoc way.

Some five months before the 1973 New Orleans gathering, at the November 1972 NCTE conference in Boston, I had made a presentation calling for a major effort to develop a bibliography of materials and establish some kind of clearinghouse for ideas and resources, arguing that one of the signs of a discipline and profession was the existence of a professional and scholarly literature. I was reacting against four common practices back then. One was the use of literary texts as reading material and as models in technical writing courses. I don't know how many times I heard people say they used descriptions from Melville's *Moby Dick* or Thoreau's observations about black ants and red ants as exemplary readings in technical writing. (At one time—depending on the makeup of the class—I had used some of those and similar passages to illustrate close observation and rendering those observations into written description, but I had quickly moved on to equally good and more relevant texts.) The second practice was assigning technical or scientific topics but requiring them to be written in the form of literary essays. The third was that almost anybody who spoke or wrote about technical writing cited John Ciardi and Jacques Barzun and the venerable Mssrs. Strunk and White as our best teachers and most influential spokesmen for clear writing. John Walter, Herman Estrin, and Tom Pearsall were beginning to be heard, but most of the professional technical writers and technical writing teachers (who were from literature or journalism backgrounds—if they were from writing backgrounds at all) tended to cite literary sources. The fourth practice was assigning technical writing classes to just about anybody who was willing to teach them or who needed to have a class to fill out a teaching load for the term. The typical teacher of technical writing at the time was only a semi-interested participant, waiting for the day that he (mostly males for sure) could teach literature courses. I guess it was only natural for English teachers to try to use what they knew best when they started teaching technical writing. What bothered me was their attitude that what they were familiar with was sufficient and there wasn't anything else worth knowing. Few of them had ever held a nonacademic job.

By 1972, I had already been a writer and editor for nearly four years for funded research projects in the old U.S. Department of Health, Education, and Welfare and the St. Louis Arsenal (U.S. Army), and I already had seven years' experience teaching technical writing and being responsible for preparing instructors and others to teach it. It struck me that those who were using Melville, Thoreau, Chaucer, and Benjamin Franklin (there was lots of excitement over the triworks, cetology, cabin building, the Astrolabe, and electrical experiments) were feeling the need to burnish the technical writing podium with a literary cloth. While interesting, these writers didn't seem to be all that relevant for the kind of writing and editing I was doing, and I certainly wasn't writing or editing literary essays. To augment my own experience in technical writing, I also read what I could find about it. Omitting the Ciardi, Barzun, and Strunk and White writings, my research produced an annotated bibliography that I co-authored with an instructor named Vivienne Hertz. (I was pushing to get all the bright females I could find interested in technical writing, for they were the ones most likely to take to it seriously. The males tended to want to hold out for teaching sophomore literature—or any literature course. They even preferred teaching composition because they could continue to teach literature as the main subject matter.) The bibliographic article was published in the May 1970 issue of *College Composition and Communication*. I no longer have the letter of acceptance, but I recall that the editor, William Irmscher, wrote back that he was pleased to receive an article on a subject that was underrepresented in the journal and that might actually be of some use to readers.

Jay Gould of Rensselaer was in that audience that day in November, 1972 when I made my presentation, and he asked me afterwards to send him a copy of my paper, which he subsequently published in the *Journal of Technical Writing and Communication*. I knew who Jay was, but I had never met him before. I believe my legs were numb when he asked for a copy of my paper. I had once spotted Robert Frost across a crowded room, and to me Jay looked a lot like him. I was mesmerized. Elated by Jay's notice, I was pumped and felt like I had just flattened somebody on a kickoff return.

With that earlier Boston presentation and the need to design activities that nurtured faculty and actually enabled some of them to begin new, interesting, and satisfying career changes in my mind, I said something at the 1973 New Orleans session that echoed my earlier presentation about the need to think about creating a clearinghouse for information about the teaching of technical writing. John Harris and I had knocked around the idea a few times before this meeting. Both of us were getting lots of queries from teachers desperate to be ready to teach the technical writing classes they had been assigned.

My long (sometimes all-night-long) talks with Harris confirmed for me that technical writing was quite different from composition taught in freshman courses. And Harris's diffuse range of knowledge and experience clarified distinctions that seemed to me to be extremely useful. Most of what I believe about technical communication—now some thirty years later—is still heavily influenced by what I learned from him.

At the 1973 CCCC session in New Orleans, it was John Harris who spoke most persuasively and eloquently about the need for an organization, and proposed that we form an organization of teachers of technical writing. Toward the end of the Q&A session, he spoke briefly about the need for an organization that would help advance the cause of technical writing. The rest, as they say, is history.

There were probably fifteen or twenty in the room, including two memorable fellows from Georgia Tech—David Comer and Ralph Spillman—who were fixtures at technical writing sessions back then. They were sharp guys and had written a textbook, *Modern Technical and Industrial Reports*, although almost everybody was using *Technical Writing*, by Gordon Mills and John Walter, or an anthology edited by Herman Estrin, *Technical and Professional Writing: A Practical Anthology*. Many of the organization's future leaders were not there, including Tom Pearsall and David Carson. Tom had just retired from the USAF and was at the University of Minnesota and becoming involved in the Society for Technical Communication. Dave, I believe, was still on active duty and was stationed in Europe.

Either Comer or Spillman (or it could have been both, because they tended to speak almost in stereo) made the formal motion to form an organization and to name John Harris as its first president. John realized that in the time-honored tradition he was going to be expected to act on his own ideas. He quickly nominated John Walter as vice president and secretary-treasurer and me as editor of the organization's journal. I attempted, in vain, to defer to Herman Estrin as editor, but Herman declined. So I accepted the nomination. That slate was elected by acclamation.

Immediately following this session, Harris, John Walter, and I met in the hallway with my department chair, who agreed to provide the funds for three years to get the journal established. It was understood that I would solicit manuscripts, set up an editorial review committee, produce the journal, and mail it out (this included typing the mailing labels and sticking them on the covers of the copies of the journal)—and I was beginning to understand Herman Estrin's desire for me, not him, to be the editor. I could not have done all those things without the help of Mary Bragg, who was in charge of institutional publications at Morehead State University at the time. Mary was indispensable in helping prepare the copy, and I owe her a special debt in helping get out the first few issues of *The Technical Writing Teacher.* We had the first issue of the journal out in six months.

That evening after the first meeting, John Harris and I did what we usually did in the free time at conferences: We walked the streets of the city—this time, those of New Orleans (not just the French Quarter), nosing around in odd shops, eateries, and ethnic groceries. Occasionally there would be a challenge to buy and eat something that neither of us recognized or had even heard of. Those were the days of cast iron stomachs and mingling with the street people. John and I dressed in clothes that didn't mark us as conventioneers, and we thought we looked tough enough that no urban mugger would regard us as easy prey. John had the remarkable knack of confounding panhandlers by hitting them up for a quarter just as they approached us. His fluent Spanish helped us find good, authentic, blue-collar places to eat in many cities of the Southwest. During those city walks, we talked a lot about technical writing—its uses and its teaching. We spent a lot of time speculating on the nature of the local architecture (I was full of John Ruskin's ideas back then) and the causes of certain structural problems we saw. I recall spending over an hour after we first checked into our hotel room in Detroit speculating on what caused the crack in the glass door of the shower. We believed, and still do, that the world is full of features and occasions for the technical writing class.

We both had similar experiences when we first taught technical writing. I had first learned that I would be teaching technical writing when I was a graduate assistant at the University of Missouri. I was helping the composition director, Will Johnson (probably the most sensible and effective administrator I ever worked for) at a drop-add session during registration. In those days, registration was held in old Brewer Field House, and the place was filled with long rows of tables (and even longer lines of students waiting to fill out forms to drop or add classes). During a brief slack moment, Will asked me how I would like teaching English 60. I confessed that I didn't know exactly what English 60 was or whether I had the qualifications to teach it. Will responded in his signature gruff way that he was sure I was qualified. When I asked him to explain the qualifications, he said that I was male, over six feet tall and weighed over two hundred pounds, a military veteran, and a former jock. In those pre–politically correct days, almost all students in technical writing were either engineering, agriculture, or forestry majors, and all were males. (It would be nearly three years before I met a woman who taught technical writing.)

In response to my question about resources, Will said, "Go see Bucco." Bucco was Marty Bucco, now a professor of English at Colorado State University. Bucco, who has had a distinguished career in twentieth-century American literature and the literature of the American West, was the senior technical writing expert in the department, having taught the course the year before, and was keeper of the "official" syllabus for the course. It didn't take long for me to see Bucco, for his desk was next to mine in the graduate assistant bullpen on the third floor of what was then the new Arts & Sciences Building. The more John Harris and I talked, the clearer it became to us that something better than "Go see Bucco" (as helpful as that actually turned out to be) was needed to prepare teachers to teach technical writing. I was intent on establishing an annual bibliography of technical writing materials, if nothing else ever came of the journal.

AFTER THAT FIRST MEETING: LAUNCHING
THE TECHNICAL WRITING TEACHER

The summer after the 1973 CCCC meeting in New Orleans, I chose the title of the journal, *The Technical Writing Teacher*, on the fly. I recalled seeing a periodical titled *The Mathematics Teacher*, and I thought that title would clearly distinguish the journal and reflect the organization's purpose. Thus, the baby was given that name. By November, 1974, we had nearly 200 members. The journal was meeting the needs of readers that the early officers of ATTW had identified. By 1980, there were almost 1,200 members. I'm not sure that we have ever regained that level. We lost a ton of teachers from community colleges and technical institutes in the mid-1980s. Many did not rejoin because they were no longer finding the journal articles as helpful to them as they once had been. The journal title, mission, content, and readership were changing in response to perceived needs for a publication that would impress tenure committees. The result, as far as I can see, is that the journal, now titled *Technical Communication Quarterly*, is one of the substantial journals in technical communication and rhetoric.

As editor, I got busy and solicited articles from John Harris, John Walter, Tom Pearsall, Nell Ann Pickett, Ann Laster, John H. Mitchell, Tom Warren, Herman Estrin, Wayne Losano, and Art Young. In the pressure to get started, I went with those I knew. Fortunately, I knew the three Johns: three of the most intelligent and generous—and most unassuming—colleagues I've ever had the privilege to work with. They knew what they were talking about, and their writing was blessedly clear. Walter's article was just what I thought was needed: a good, clear definition of what technical writing was all about, how he got into it, and what he believed made for an effective technical writing course. And it ended with a self-effacing statement saying that he (even with all his experience) had no conviction that he was doing the best that could be done and was looking forward to hearing how others dealt with the problems of teaching a basic course in technical writing.

Having done bibliographical work, I knew the pitfalls of putting one together, so I hastily created some guidelines for coverage and consistent formatting, and we had an annual bibliography out quickly. The first bibliography committee consisted of a few people that I knew I could rely on, and, of course, most were female: Dorothy Bankston, Carolyn Miller, and Vivienne Hertz. Tom Warren and I were the men on the committee. The first annual ATTW bibliography appeared in the Fall, 1975 issue of *The Technical Writing Teacher.*

THE NEXT FEW YEARS: DEVELOPING RELATIONSHIPS WITH NCTE AND CCCC

I think I've described most of what happened at that first meeting in New Orleans. I don't recall that ATTW as an organized group had any formal meetings at CCCC or NCTE. However, the group began piggybacking on the activities of the NCTE Committee on Technical and Scientific Writing, which met at both the NCTEcon in November (with large audiences, until NCTE made a big shift toward secondary school sessions) and at the CCCC in the spring. I think ATTW began meeting officially and as a separate group at CCCC about the time CCCC instituted special interest group venues. I can't remember when those began, although I believe the first special interest group meeting was in New York in the early 1980s.

A lot of ATTW business was conducted at the end of meetings of the NCTE Committee on Technical and Scientific Writing. (The CCCC Committee did not exist back then, and to this day, I'm not sure why it was formed or why we have two committees. However, the more representation and activity we have, the better.)

With ATTW established, John Harris, Herman Estrin, John Walter, Nell Ann Pickett, Virginia Book, I, and others desired more focus on technical writing needs than what the NCTE or CCCC or MLA were able or willing to provide at the time. At this time, John Walter, Tom Warren, and Tom Pearsall were making similar inroads in the Society for Technical Communication, and the latter two were working to establish the Council for Programs in Technical and Scientific Communication (CPTSC). CPTSC was founded to concentrate on developing and maintaining professional programs and give directors of such programs a venue for interaction. Some felt that ATTW and the NCTE Committee were concerned primarily with helping prepare teachers for the introductory technical writing course. I was pleased to see so many groups forming to advance the cause of technical writing: ATTW, the NCTE Committee, the CCCC Committee, CPTSC. Whether they have worked as well together as they could have is a topic for another time.

ATTW and the NCTE Committee on Scientific and Technical Writing had virtually the same officers. Herman Estrin and I were serving as chair and associate chair of the NCTE Committee on Scientific and Technical Writing (as it was called then), and we were succeeding in getting both NCTE and CCCC to provide us with more time on their convention programs. We had been given a couple of sessions each year, and those sessions were always well attended, even though they were not scheduled at particularly desirable times. All that was to change in just a year or two. One of the things that lit a fire under the NCTE was the information I passed on to NCTE folk.

In November, 1974, Herman was a member of the NCTE Executive Committee, but he was unable to attend the meeting and designated me to represent him at the annual meeting. At the time, I did not know there was such a thing as an executive committee and had no idea what it did. At that meeting, I—perhaps inappropriately and too aggressively—spoke about technical writing and what ATTW was beginning to do. I don't recall getting much more than a polite hearing. However, things quickly changed after that convention, primarily as a result of what we did in the meeting of NCTE Committee on Technical and Scientific Writing at that convention. I chaired that meeting (Herman was ill) and Nell Ann Pickett wrote the minutes. We laid out an ambitious plan to have more sessions of technical writing at the conference and proposed a book for NTCE publication.

John Maxwell, then the Deputy Executive Director of NCTE, wrote me a brief response dated January 31, 1975 to those minutes, stating, "Yours must be the best set of minutes that I have ever received from a Council committee." When I spoke with him by phone later, I asked him rhetorically what else he thought he would get from our committee—it's our business to conduct meetings and write well. Maxwell and Sister Rosemary Winkeljohann, Director of Member Services for NCTE, told me that they were aware of our needs in technical writing and began adding sessions at both CCCC and NCTEcon for us. Later that year, NCTE published a collection of essays on teaching technical writing that was edited by Herman and me (*The Teaching of Technical Writing*, 1975). Paul O'Dea, who was director of publications at NCTE at the time, told me that the book was one of the best they had published in several years and was selling well. (It wasn't until a few years later when I served on the CCCC Executive Committee that I learned how skimpy the sales were for CCCC books, including books by well-known rhetoricians who were influencing the discipline.) Winkeljohann wrote to me about the praise she had received from a standing-room-only crowd that had attended a three-day precon workshop in technical writing in San Diego in November, 1975. That workshop—the third three-day technical writing precon workshop in three years—was conducted by Herman Estrin, John Harris, Nell Ann Pickett, John Walter, and me. We were all members of the NCTE Committee, and each one except Herman was an officer in ATTW. We were on our way then. John Harris, Herman Estrin, John Walter, and I were practicing the old Cowbird principle of using the NCTE and CCCC for technical writing purposes.

In those crucial first years, we received much kindness and help from several NCTE and CCCC leaders. For several years during the mid- to late 1970s, the officers of ATTW and the members of the NCTE Committee on Technical and Scientific Writing provided advice to conference chairs and planners for sessions at NCTEcon and CCCC. From 1974 to about 1981, the number of sessions increased as the attendance grew. And NCTE—pleased with the reception and sales of *The Teaching of Technical Writing*—had two more books on technical writing in the hopper.

Another major success of ATTW and the NCTE Committee came in the late 1970s when Dave Carson succeeded in persuading the leaders of NCTE and its Executive Committee to establish the publication awards in technical and scientific writing. These were, I believe, the first publication awards ever to carry the NCTE imprimatur. What an acknowledgement we owe Dave for that one!

As chair of the committee, my first experience with minor turbulence came with an incoming CCCC program chair in 1980 or 1981 over the lack of minorities (specifically African Americans) as speakers in technical writing sessions. Fortunately for us, Pearl Saunders of St. Louis was aboard by then, but we had to admit that we were not as successful as we would like to have been in identifying minority participants. In response to my challenge, the program chair that year backed off when she was unable to identify another minority teacher of technical writing who was a member of NCTE or CCCC. She accepted our committee's full slate of sessions and speakers. This conference, I believe, still holds the record for the number of separate sessions on technical writing at CCCC.

The following year, the number of sessions dropped considerably. The program chair that year rebuffed my attempts to assist. He informed me that he felt absolutely no need for help from ATTW or the NCTE Committee on Technical and Scientific Writing, for his wife taught technical writing and she was all the assistance he needed.

For several years, the Executive Committee of ATTW and the NCTE Committee on Scientific and Technical Writing/Communication discussed the feasibility of establishing ATTW's own annual conference. Through the 1970s and early 1980s, annual ATTW meetings found a convenient place at the CCCC—less so at NCTEcon after that group began emphasizing secondary teachers. However, it took nearly another fifteen years before Sam Dragga, Carolyn Rude, and others had the sensible—and great—idea to have a dedicated time for ATTW at the beginning of the CCCC conferences and succeeded in selling it to the leaders of the CCCC. The organization has changed over the years, and the annual ATTW sessions at the CCCC conferences are managed and organized much better than when the paper and sessions proposals had to be subjected to decisions by others who, as the years passed, knew or cared very little about technical communication.

CONCLUSION

Because this account is unrehearsed and quickly written, I have no strategy for ending. But this has gone on too long, so I will end with this paragraph. Lots of hard work by a few people led to gradual and cumulative progress. The work took patience and endurance as we laid the bricks for the foundation of ATTW one at a time. There were some paradigm breaking moments, and I've tried to identify most of them. Ultimately the building and growth of ATTW is fundamentally a social undertaking, and there are many who have contributed to the life of this organization. ATTW evolves as individuals identify needs of people in similar circumstances and implement plans to meet those needs. The organization depends on such individuals, but when the individuals work together to meet perceived needs in the field, they accomplish so much more than they could working alone.

WORKS CITED

Comer, David B., and Ralph R. Spillman. *Modern Technical and Industrial Reports.* New York: Putnam, 1962.

Cunningham, Donald H., and Herman A. Estrin, eds. *The Teaching of Technical Writing.* Urbana, IL: National Council of Teachers of English, 1975.

Estrin, Herman A., ed. *Technical and Professional Writing: A Practical Anthology.* New York: Harcourt, 1963.

Mills, Gordon H., and John A. Walter. *Technical Writing.* New York: Rinehart, 1954.

Donald H. Cunningham is professor of English at Auburn University and director of its program in technical and professional communication. He served as editor of *The Technical Writing Teacher* for eleven years, from 1973 to 1984, and was a founding member of ATTW. He is a fellow of ATTW and STC and has won STC's Jay R. Gould Award for Excellence in Teaching Technical Communication. Four of his books have won awards from the Society for Technical Communication and the National Council of Teachers of English.

REVIEWS

Tracy Bridgeford, University of Nebraska at Omaha, Editor

Reshaping Technical Communication: New Directions and Challenges for the 21st Century. Ed. Barbara Mirel and Rachel Spilka. Mahwah, NJ: Lawrence Erlbaum Associates, Inc. 216 pages.

Reviewed by Tracy Bridgeford
University of Nebraska at Omaha

Good research, like a good story, is not discovered; it is interpreted, analyzed, shaped, and reshaped so that members of a community understand and accept its value. And like a good story, *Reshaping Technical Communication*, demonstrates a level of verisimilitude—or tellability—that convincingly presents a collective history, adds value to that history, and negotiates the future status of the technical communication community. What makes this collection tellable is the comprehensiveness with which the editors and contributors portray technical communication as both "resting on its laurels" and "reinvigorating" its collective history (1–2). According to the editors, this collection aims to "reinvigorat[e] our status, identity and value" by reshaping the perspectives, methods, and stories we tell about our discipline (2). Its value as a collection for researchers, teachers, students, and practitioners is its expansive view of the history of academic-industry relationships.

Part of the editorial framework, however, lacks verisimilitude in that the editors identify the need for reinvigoration as necessary because the field has not "kept pace with either the transformations wrought by the technologies with which we work or the growing demands for effective, valuable, and satisfying interactions with technologies and information systems" (2). This technological preference provides the groundwork for what the editors identify as the problem with current research: the status of technical communication, technical communication researchers, and technical communicators. The question of status in technical communication (for academics and practitioners) is a story that has been retold vigorously with each decade since the 1970s. From the instrumental vs. rhetorical, skill vs. knowledge, or humanistic vs. technical perspectives, the question of status has permeated our research discussions as we struggle to identify who we are and our value to the world (see, e.g., Moore; Miller, "Humanistic"; Dombrowski; and Johnson). Although the story of status is something that has always been a problem in our field, as the editors suggest, the retelling of this story does not reinvigorate history as much as it solidifies the "identity crisis" with which we are already very familiar (4).

The book is divided into two parts, which together explore the two sides of the academic-industry relationships. In Part I, "Revising Industry and Academia: Cultures and Relations," the contributors explore current relationships between industry and academic factions, proposing "conditions for accommodating more dynamic and flexible academic and industry contributions to training, research, and practice" (8). These conditions include identifying cultural differences (Dicks), exploring overlapping spaces between industry and academia (Blakeslee), and creating communities of practice as a bridge for active-practice (Bernhardt). Although these chapters each basically tell the same story about past efforts to build relationships between academia and industry, they each provide different parts of the putative whole. Dicks, for example, broadly identifies the cultural differences between industry and academia, while the rest of the Part I contributors explore more discrete parts, such as exploring the overlapping spaces that join these cultures (Blakeslee) or the community of practices academics and practitioners could create by working together (Bernhardt).

In Part II, "Re-envisioning the Profession," the authors investigate topics such as the credibility issues associated with becoming a profession (Spilka), the problem with dissemination of research in information design (Schriver), the need for a corporate-university hybrid sort of learning (Faber, Johnson-Eilola), the expansion of support for career transitions (Anscheutz and Rosenbaum), and the role of usability specialists for complex activities in context (Mirel). In some ways, the editors and contributors realize the promise of the collection's title—reshaping technical communication—by providing what the editors call a "more expansive vision of what we could or should become" (4) and by reemphasizing the strategic value of "what we add to our work contexts and product designs and directions" (92). In stressing this valued added, however, the contributors call more attention to the need for status than the actual status.

For a story to be worth telling, it must be about how a familiar, conventional, story has been breached in some way that questions or reverses that story (Bruner). From this perspective, two chapters in particular provide such a breach by revealing and interpreting two significant ideological blind spots in our history.

In her chapter, Deborah Bosley explores what she sees as three barriers that stand in the way of academics forming collaborative relationships with industry: Academics tend to focus on differences rather than on similarities (see Dicks's chapter for this discussion), to distance themselves from practitioners in unproductive ways, and to underestimate the value of academic research to practitioners. From a perspective of identity, Bosley explains that academics must learn to define themselves as practitioners in order to learn more about the work environment" (35) and "to publish the results in forms and publications that practitioners read" (38). Given the early battles for academic respect, which probably continue yet in some departments across the country, I imagine that some members of the technical communication community might cringe at the term *practitioner* because early efforts to identify technical communication as a bona fide scholarly endeavor suffered from our association with practical skills (see, e.g., Miller, "Practical").

Anthony Paré's perspective of "writing in the service of something else" (59) could cause a similar response. Like practitioner, the term *service* has a pejorative history in our field because it is associated with paying one's dues (i.e., until literature courses became available) or providing mere support for other fields rather than a field of study with its own intrinsic knowledge and value. But Paré's focus on writing as "guided expertise" that transforms over time (as opposed to one-time classroom testing) insightfully reminds us that practitioners use "writing to perform or participate" (62) in workplace activities or actions (what Bernhardt in this volume refers to as *active-practice*). The breach in Paré's version—writing as performance or participation—reveals how often, no matter how much we protest, we talk about writing comprehensively in terms of assignments, research, or assessment, but not as part of other activities. We forget that we choose to focus our research on writing as an area of study, but practitioners use writing to do something in particular with the writing.

For both Bosley and Paré, practitioners might find academic research more valuable if it existed in a form designed with the knowledge that time and relevance influence a practitioner's decision to read an article or a book that doesn't directly connect with the exigency of their work. As academics, we sometimes forget how demanding workplace schedules and deadlines can be (which is not to say we don't experience similar grueling demands) or how production mishaps often hinder practitioners' progress and efficiency.

Of special interest to me is the last chapter: Russell Borland's "Tales of Brave Ulysses." To fully appreciate the creativity with which Borland sets forth his chapter, readers should take the time to read Thomas King's *Green Grass, Running Water.* King's novel is a masterful storytelling experience in which he reshapes some hard-held Christian myths through the trickster figure Coyote. Borland's chapter demonstrates the depth with which technical communicators understand themselves, their identities, and their value to organizations. In this historical narrative, Borland captures the essence of the stories we tell about who we are. In fact, Borland demonstrates, just as Thomas King does, the danger of retelling the same story, such as ours about status, without harmonizing it with current circumstances and audiences.

Because this collection provides insight into the value question, it should be included on every reading list for comprehensive exams, course research seminars, and dissertation proposals. Current and future PhD students, researchers, and dissertation directors might begin their explorations by reviewing the research agenda outlined in the Appendix. I'd suggest, though, that when developing research questions, researchers begin with the assumption that we are valuable, not work their way toward it.

Good research, like good stories, should challenge readers to see what is familiar while providing new interpretations. The editors and contributors to this volume certainly provide a plethora of venues for crafting future stories about technical communication. If anything is missing for me, it is direct reference to pedagogical research, but perhaps that topic warrants a second volume. This collection offers mostly what its editors promise: "to propose far-reaching, innovative, and nontraditional strategies, visions, and ways of thinking" (5). Its overall purpose—to "inspire readers to continue the discussions and debates introduced here" (5–6)—challenges each of us to participate more fully in the story of technical communication in ways that add value to our knowledge domain.

WORKS CITED

Bruner, Jerome. "The Narrative Construction of Reality." *Critical Inquiry* 18 (1991): 1–21.

Dombrowki, Paul M. *Humanistic Aspects of Technical Communication.* New York: Baywood, 1994.

Johnson, Robert R. *User-Centered Technology: A Rhetorical Theory for Computers and Other Mundane Artifacts.* New York: State University of New York Press, 1998.

Miller, Carolyn R. "A Humanistic Rationale for Technical Writing." *College English* 40 (1979): 610–17.

Miller, Carolyn R. "What's Practical About Technical Writing?" *Technical Writing: Theory and Practice.* Ed. Bertie E. Fearing and W. Keats Sparrow. New York: MLA, 1988. 14–24.

Moore, Patrick. Instrumental discourse is as humanistic as rhetoric. *Journal of Business and Technical Communication* 10(1) (1996): 100–18.

Printed in the United States
by Baker & Taylor Publisher Services